Aircraft
Partnership

Other McGraw-Hill Aviation Titles

Aircraft
Partnership

Geza Szurovy

McGraw-Hill

New York   San Francisco   Washington, D.C.   Auckland   Bogotá
Caracas   Lisbon   London   Madrid   Mexico City   Milan
Montreal   New Delhi   San Juan   Singapore
Sydney   Tokyo   Toronto

Library of Congress Catalog Card Number 97-75975

McGraw-Hill

A Division of The McGraw·Hill Companies

5 6 7 8 9   BKM BKM   0 9 8 7

ISBN 0-07-063347-9

Cover and all photos by Geza Szurovy unless stated otherwise. The sponsoring editor for this book was Shelley Ingram Carr, the editing supervisor was Paul R. Sobel, and the production supervisor was Clare B. Stanley. It was set in Garamond in the GEN1-AV2 design by Kim Sheran and Michele Bettermann of McGraw-Hill's Professional Book Group composition unit, Hightstown, N.J.

McGraw-Hill books are available at special quantity discounts to use as premiums and sales promotions, or for use in corporate training programs. For more information, please write to the Director of Special Sales, McGraw-Hill, 11 West 19th Street, New York, NY 10011. Or contact your local bookstore.

Contents

Contents

Introduction

Without a doubt an aircraft partnership is the most effective way to attain the goal of affordable, high quality flying. Entering into a partnership of even two pilots dramatically reduces the costs of flying. It cuts the cost of buying an aircraft in half and slashes the cost of keeping and flying it compared to going it alone. If you fly it enough it also handsomely beats the rental alternative.

This book is your A to Z guide to evaluating, organizing and operating an aircraft partnership. It covers the topic of aircraft ownership from the unique perspective of partnerships and is an equally suitable primary resource for both the novice and the experienced aircraft owner seeking the highest quality of flying for the least expense.

The roots of this book are in an earlier work of mine, *Fly for Less*, which addressed both partnerships and flying clubs. The overwhelming interest expressed in aircraft partnerships prompted by *Fly for Less* induced me to address this topic on its own in much greater detail.

Chapter 1 makes the case for the joint ownership of aircraft over individual ownership and the rental option. It provides guidance on how to develop the right attitude to evaluating your alternatives and setting your sights on the optimal solution. Chapter 2 discusses the importance of partner compatibility. It offers advice on how to identify the partners who are right for you and provides guidance on setting partnership goals. Chapters 3 and 4 discuss in great detail how to analyze the costs of flying and match your partnership opportunities to your flying budget.

Chapter 5 is a comprehensive discussion of the co-ownership agreement, that all important document which spells out with great precision how the partnership will function. A good co-ownership agreement up front is the best preventative medicine for massive headaches down the road. Sample co-ownership agreements are presented, and the question of incorporation is discussed.

Chapters 6 and 7 cover financing and insurance, starting with the basics and moving on to the particular borrowing and insurance issues faced by the partnership. You will learn about alternatives to bank financing and how to get the best insurance package for your group.

Chapter 8 covers the day to day operations of the partnerships. It shows you how to set up a suitable scheduling system, how to manage the finances of your group, including collecting payments due from partners and paying the bills, and how to take good care of your airplane.

Chapter 9 discusses the pros and cons of partnerships of various size and encourages you to consider creative partnerships, such as a partnership to learn to fly and one to build an experimental airplane. Chapter 10 addresses the special case of the pilot joining an already existing partnership, pointing out the advantages as well as the potential pitfalls.

If a partnership grows, it eventually becomes big enough to metamorphose into a flying club. Chapter 11 covers this transition and provides a summary discussion of flying clubs.

Chapter 12 provides the last bit of crucial advice required to realize your partnership goals. It is a primer on buying an airplane. It will take you from identifying potential aircraft through evaluating them and closing the deal on the right one.

The Appendixes provide you the tools to set up your own aircraft partnership. Here you will find data gathering forms, financial analysis worksheets, the layout and formulas required to build your own Excel financial analysis templates, sample co-ownership agreements, aircraft record keeping forms, checklists for buying an airplane, and more.

Buying and flying airplanes in partnerships is integral to revitalizing general aviation. For many of us an airplane partnership will increasingly make the difference between continuing to fly or grounding ourselves. So let's recognize at last the great joint ownership opportunities that have been out there for years, and keep 'em flying.

I would also like to take the opportunity to thank the many pilots who are already converted to the cause and shared their experience with me. I would particularly like to acknowledge the contribution of Herb Hill, Ford Rackemann (who is flying several of the aircraft in the photographs), Gary Arber, Jerry Owens, The Mid Island Air Service, and all the pilots who responded to my column on affordable ownership in Private Pilot, especially Augustus Brown, Jerry Brown, and Margaret Bickers.

Geza Szurovy

1

Why an Aircraft Partnership?

A hot trend in the ownership of aircraft is the fractional ownership of executive jets. In an age when getting the most out of assets is becoming a basic doctrine of corporate culture, the captains of industry are finding it unjustifiable to buy a Citation for $5 million if they need to fly it for only 100 hours a year. Instead, they opt to buy only a one-eighth share in the Citation for $625,000 and fly it the same 100 hours a year that they would if they owned the entire jet. The same amount of flying in the same airplane for $625,000 instead of $5 million made possible by joint ownership; that should please even the most irascible shareholder.

The savings are equally dramatic when the aircraft being purchased is a light airplane. A new, well-equipped Cessna 172 costs $130,000, a price that raises howls of complaint from the pilot community. The fact is, however, that given the small volumes in which light aircraft are produced today, the manufacturers are hard put to charge any less and still make enough money to stay in the light airplane business. So what to do if you don't have that kind of money, or, even if you do, you think it is too much for the amount of flying you plan to do? Find three other like-minded partners.

The four of you can buy the new $130,000 airplane for $32,500 each, which is about as much most people are willing to fork over for their favorite car or sport utility vehicle. If you are willing to be even slightly flexible, you should have ample airplane availability to meet your flying needs (captains of industry, who have less flexibility in scheduling travel, ensure availability by buying their aircraft share from an airplane management company that maintains an entire fleet

Fig. 1-1. *The Ercoupe is an ideal low-cost classic.*

of identical jets, all fractionally owned and available to all owners in the program).

The numbers get even better if you look for a good used airplane with the same performance as the new Cessna 172. Although used aircraft prices are steadily rising, there are still plenty of used four seaters with 172 performance out there in good condition that can be had for around $40,000. With four partners that is only $10,000 per partner. You could put that on a credit card!

Let's take a closer look at some of the advantages of an aircraft partnership. The benefits depend on the alternatives to which you compare the partnership. If single ownership is the other choice you are considering, a partnership is a far less costly option. If you are willing to accommodate to some sort of a scheduling system, the chances are excellent that you'll accomplish all the flying you set out to do for far less than it would cost to reach the same goals owning alone.

For a given amount of flying, the smaller the partnership, the greater the aircraft availability. It is often said that a two-person partnership is practically like owning an airplane alone. Three or four partners can also comfortably manage availability, and if you are mostly into short, local hops, even larger partnerships work out very well.

For many pilots a partnership may be the only avenue to aircraft ownership because they don't have the financial resources to own an airplane alone. But there are other financial benefits to a partnership compared to single ownership that go beyond buying the airplane.

By sharing such costs as hangaring or tie-down, insurance, maintenance, annual inspections, the addition of avionics and equipment, and more, your per partner costs are further reduced.

Partnerships can also benefit pilots who can afford to own alone. By choosing a partnership, these pilots can afford a more capable, more expensive airplane than if they remained on their own. As you will see, in a three-person partnership you can own and fly a Bonanza 100 hours per year for what it would cost you to own and fly the same amount of time in a Piper Archer as its sole owner.

Beyond a certain number of flying hours partnerships also compare favorably to the cost of aircraft rental. The per hour operating costs (fuel, oil, etc.) of the partnership are lower than the rental rate for a comparable aircraft. The partnership's fixed costs (hangar, insurance, registration fees, etc.) per flying hour are a function of the hours flown, and below a minimum number of hours can exceed the rental rate, but as the number of hours increases the costs steadily fall. It doesn't take too many partnership flying hours to reduce the hourly rate to below the comparable rental rate.

The nonfinancial benefits of the partnership easily outweigh the rental option. You do, after all, own your own airplane, even if it is with co-owners. You are competing for access with fewer pilots than the customers of an FBO (Fixed Base Operation), and face fewer restrictions on length of use and no daily minimums.

Fig. 1-2. *The Piper Turbo Arrow may be suitable for partners who travel.*

As the costs of flying increase we are all being forced to accomplish more with less, to be more efficient in our flying. To keep general aviation from shriveling into the amusement of the rich we need to do everything in our power to keep as many pilots flying as possible. The Young Eagles program, cajoling the legal system and politicians to establish a reasonable liability environment, and encouraging manufacturers to be innovative are all worthwhile efforts. But the airplane partnership, the fix most readily available that rolls back the cost of flying to levels thought gone forever, should also be enthusiastically embraced.

According to Federal Aviation Administration (FAA) statistics, the average privately owned light airplane in the United States flies 50 hours or less per year. That is why, when a friend of mine goes out to the airport and surveys the ramp, he says, "I don't see airplanes. I see underutilized assets." There is tremendous underutilized capacity in our general aviation fleet. It is time to stop complaining and join together in airplane partnerships to show up the high cost of flying for the myth that it is.

Start with the Right Attitude

When you begin to consider an airplane partnership, start with a clean slate and focus on your flying objectives first. What kind of flying do you want to do, in what type of airplane, and do you have or can you get the skills that it will take? Consider your financial resources only after you've clarified your flying objectives. Then do a detailed analysis of all the partnership options to see what kind of partnership would comfortably meet your goals.

It is important not to impose any constraints on your thinking up front because of your financial situation, because you never know what the analysis will show you until you get into it. Only then will you begin to see options and alternatives to close any gaps between the partnership structure and your financial resources. And always think "How can I make this work?" rather than "Why won't this work?"

Buy the Aircraft That Best Meets Your Needs

Another thought to keep in mind as you contemplate a partnership is to choose the airplane that best meets your needs. No more, no less. If you settle for a less capable airplane you will not be able to accomplish the flying objectives you set for yourself and will ultimately become frustrated by flying. A good example of this situation is buying an airplane with no autopilot when you plan to fly a lot of single-pilot IFR (instrument flight rules).

Overbuying can be equally counterproductive. Hauling out a Cessna P-210 equipped with all the avionics known to civilization to hop over to the coffee shop 60 miles down the airway will hardly seem worthwhile after the first few times. It is discouraging to end up with a supership in the hangar and have to admit to yourself that low and slow CAVU VFR (ceiling and visibility unlimited visual flight rules) is what you really like.

Your skill levels and the time and resources you have to improve and maintain them is another important consideration when you choosing an airplane. If you overbuy for your current qualifications and are unable to upgrade your skills (for whatever reason), you'll end up being afraid of the airplane and every flight will be a potentially dangerous ordeal.

Don't Stretch Yourself Financially

It is tempting to scrape together every last penny to pull off the partnership of your dreams by a financial hair. It is also a mistake. If you

Fig. 1-3. *The Mooney Ovation is ready for business or pleasure.*

can't comfortably afford the partnership you'll end up flying less and sweating every expense. Each year will be a potential financial disaster, and if you do get caught short and get behind on the airplane bills you'll also be letting down your partners. It is much better for all concerned if each partner has comfortable financial wiggle room to be able to easily take care of unexpected expenses.

Avoiding financial strain does not necessarily mean that you have to give up on the type of aircraft and flying you desire and settle for less. Adding another partner or two when you set up the partnership can put you right back in business. A Pitts S-2B that is marginally attainable for two may be easily affordable for four.

Select the Right Partners

No matter what the numbers tell you and how perfect your choice of airplane is, if you end up with the wrong partners your partnership dream can quickly turn into a nightmare. Not having the right partners is the main reason for the unraveling of aircraft partnerships. Never lose sight of the fact that the right partners are key to the success of the partnership. This point is so important that before getting into the technical details of forming an aircraft partnership I'll devote the next chapter to selecting the right partners and getting along with them.

Case Study: Cessna 172B–Three Partners

The aircraft

This 1961 Cessna 172B is an Avcon conversion equipped with a 180-hp Lycoming engine and a constant speed propeller. It also has two wing fences and drooping wingtips to enhance short field performance. It has been the partnership's airplane for 5 years and was quite a find with only 1900 hours on the airframe and 400 hours on the engine since major overhaul.

Partnership background

This partnership is a three-way partnership. Two of the partners, Herb and Frank, started the partnership 11 years ago in another, less powerful Cessna 172. Five years ago they decided to get a more powerful airplane because they fly in and out of short fields and wanted

Case 1-1. *The Cessna 172 is one of the most popular airplanes ever made.*

the extra performance. When they bought the current airplane they took in Stan, the third partner, to help defray the extra expenses.

The airplane has basic IFR equipment but is used exclusively for VFR pleasure flying. Herb also delivers the *Atlantic Flyer,* a monthly East Coast aviation newspaper, with it to area airports. The airplane is based at a 2800-foot local airstrip.

Herb and Frank, who live in the same town, knew each other socially before deciding to form the partnership. Herb was flying a Cessna 150 at the time, an arrangement that was about to cease, and Frank was inactive but motivated to get back into flying. Neither was interested in spending the money to own an airplane alone and both found rentals too expensive, so they decided to give the partnership route a try.

Stan, the third partner, was introduced into the partnership by Frank, who knew Stan socially. He had relatively little experience at the time, but had done some ambitious national cross-country flying which Herb and Frank took as a good recommendation. Stan flew with both partners and they got to know each other better as all three participated in looking for the new airplane before formalizing the three-way partnership.

The three partners each fly approximately 50 hours or less a year.

Partnership structure

The partnership is technically an unincorporated co-ownership. It is governed by a written partnership agreement. The partners met and hashed out all the ideas they had for operating the partnership, which Frank turned into a written agreement. Under an informal understanding Herb manages the maintenance and Frank keeps the records.

Finances

All three partners paid an equal share for the airplane. No money was borrowed by the partnership to acquire the airplane. For the last year the expenses have been as follows (expense items will be explained in detail in subsequent chapters):

ANNUAL FIXED EXPENSES ($)		(3 Partners)
	Total	Per Partner
Tiedown/Hangar	65	22
Insurance	635	212
State Fees	125	42
Annual	750	250
Maintenance	2,250	750
Loan Payments	0	
Total Fixed Expenses	3,825	1,275

HOURLY OPERATING EXPENSES ($)

Total Hourly Op. Exp.		38
$/hour @ 50 hr/yr	(fixed exp/50)+op exp	63
$/hour @ 100 hr/yr	(fixed exp/100)+op exp	51

Maintenance expenses proved to be heavier than the historic average for the past 2 years in a row because of a lot of small things that needed labor-intensive work, such as restoring chafed baffling.

The fixed expenses are divided equally among the partners. Everyone pays his own operating expenses (fuel oil and engine and maintenance reserves per hour). Each partner has his own separate fuel account at the airport. Everyone tracks the hours flown in a book kept in the airplane. This information is supplied to Frank who keeps the records.

Settlement of the expenses is once a year. This is less frequent than generally recommended, but it works for the partnership because of each partner's conscientiousness. An airplane account is maintained at a local bank where the paid-in reserves are kept. However, checks are not written on this account to make payments for airplane expenses. Usually the partner handling the bill pays it personally and is then reimbursed from the airplane account. If a large extraordinary payment is required during the year an assessment is made. With one minor exception of a small delay on one fairly large payment due, this system has worked well.

Operations

With only three partners and schedules that do not conflict, operations are informal and free of friction. Herb flies only during the week, Frank only on weekends, and Stan has been quite inactive recently. When a partner wants to fly he calls the other two, preferably with a day's notice. Conflicts are resolved informally.

A flight log is kept in the glove compartment in which the date of the flight, duration of the flight (tachometer time), the name of the pilot, and total tachometer time are recorded. Each partner has a separate fuel account and the airplane is left fully fueled after each flight.

Maintenance

Maintenance is done by the local maintenance shop. The partners do not do any of their own work on the airplane, partly because the shop discourages it.

Herb acts as liaison with the maintenance shop. Every repair item other than routine oil changes is first discussed by all three partners before Herb gives the shop the go-ahead. Any major upgrade in equipment is decided on by consensus.

The partnership experience

The partnership experience has been unequivocally positive for all three partners. Each has found the others to be flexible, thorough, and trustworthy. Financially it has also been a success. None of the partners would have owned an airplane alone because based on usage it would not have made financial sense. In comparison to rentals there have been times for each partner when, due to the low usage of the airplane, renting would have been marginally less expensive, but the flexibility of ownership has outweighed the extra expense on these occasions.

2

Partner Compatibility and Goals

Entering into an airplane partnership is like getting married. You have to like each other and, equally important, you have to have reason to believe that you will go on liking each other. For a long time. Furthermore, this belief has to be mutual. Having said that, there are marriages where each day is the Fourth of July and there are others where each day is Monday morning. Most end up somewhere in between. At the end of the day a marriage will endure only if the fundamentals were there to begin with. The same goes for an airplane partnership.

The Importance of Personal Compatibility

The single most important factor in the success or failure of an airplane partnership is the partners' personal compatibility. It ranks far above the suitability of the airplane, the right price, and the size of the prospective partners' wallets. You can make no greater mistake than to be seduced into a partnership with incompatible partners by the airplane of your dreams and the financial advantages of the partnership.

Some difference of opinion is bound to arise from time to time, even among the best of friends. If the relationship is strong, the issues are quickly aired and sorted out to mutual satisfaction. If, however, the partners are incompatible, sooner or later some form of conflict will develop that will prove to be not only downright nasty, but irreconcilable.

Fig. 2-1. *The Cessna P-210 needs partners with similar goals and flying qualifications.*

In the workplace much is made of the importance of learning to get along. You can't always choose your coworkers, and presumably you want to stick around because you enjoy or need your job, so the incentive to make every effort to coexist is great.

When you are setting up an aircraft partnership you have greater flexibility. No potential partner is being forced by some higher authority to coexist with someone not of their choosing. You can and should be much more choosy about who your "colleagues" will be. If there are potential partners with whom you don't get along, don't even bother to try. You don't need the aggravation and can afford to say no.

The problem is that many of us have a hard time determining up front whether we'll have problems with someone once we get into a partnership with them (just like some marriages). Knowing somebody casually tells us little about how they'll handle a partnership, and sizing up strangers with whom you come together only because you are looking for airplane partners is even more difficult. To minimize the chances of making a big mistake it is in the interest of all potential partners to carefully and systematically assess how compatible they are likely to be in the partnership.

A good way accomplish this objective is to openly discuss all the issues of putting together and operating the partnership, and then begin to plan its structure and conduct the financial and operating analyses of potential aircraft types. Clearly articulate what each of you wants out of the partnership. Pay particular attention to any points of potential friction and open conflict and go through together in great detail how each of you would resolve these conflicts.

During the course of this exercise you'll each have ample opportunity to assess each other's personality traits and evaluate the compatibility of your respective goals and values. Before you collectively decide to take the plunge, be sure to go flying with each other. Behavior in the cockpit is revealing about compatibility in attitudes toward flying and treatment of an airplane, but bear in mind that the ability to fly well is not linked to any particular personality type.

It is in everyone's best interest to be totally honest with each other as well as themselves as you collectively go through this process. Be yourself. Tell your potential partners how you really feel about them and the issues, and expect them to do the same. If you say what you think they would like to hear or what you think is expected by social conventions even though your own opinion is different, you are asking for trouble.

Before you start scrutinizing your potential partners, look in the mirror. Do you consider yourself partnership material? Can you think of the last time you met someone halfway in a disagreement and were satisfied with the outcome? Do you feel resentment if something isn't going your way but have difficulty in initiating an open discussion of the issue? Do you hold a grudge when you don't get your way? Do you still feel a swell of anger or bitterness when you think of something unpleasant that someone did to you 10 years ago? Depending on your personality, an airplane partnership may be less than ideal for you. You can save yourself and others a lot of trouble by assessing your personality as critically as those of your potential partners.

Assessing Personality Traits

The complexity of human nature makes it difficult to accurately group people by highly detailed behavioral characteristics. However, it is possible to get a general idea of what broadly defined categories potential partners seem to fall into on the spectrum of personality traits, thus providing useful clues about how well they are likely to get along.

A partnership is a team, expected to work together in harmony. We can therefore rely on some basic concepts commonly used to study team behavior to get an indication of the potential partners' personality traits and their suitability to join together a partnership.

An important element of a team's ability to work well together is the ability of its members to communicate effectively with each other. How well a team member communicates is a good indicator of that member's personality traits and how well he or she fits into the team. One useful method of evaluating team communication breaks down communications traits into four categories and assesses where each team member fits. The categories are aggressiveness, assertiveness, responsiveness, and nonassertiveness. If during this preliminary stage of investigating the partnership options all the team members honestly communicate in a style that reflects their natural inclinations, the categories offer reliable indications of each member's underlying personality traits.

Aggressiveness

Aggressive communicators are self-centered and intolerant. Theirs is the "I, me, and myself" style. They have little use for the opinions and rights of others and are poor listeners. They believe they have all the answers and are intolerant of alternatives proposed by others. They want control and frequently attempt to exert it by putting down alternative opinions offered by others. The put-down can be subtle, especially in the initial stages of a relationship, so careful attention has to be paid to exactly what their final position is on a conflicting issue.

Aggressive types are obviously poor candidates for such a cooperative venture as an airplane partnership. As long as everything goes their way they can function effectively, but as soon as something is not to their liking they show signs of intolerance.

The era of political correctness has put aggressive types under considerable pressure, and many have become quite clever at masking their true personalities when dealing directly with others on an issue important to them, such as putting together an airplane partnership. Once the partnership is established, unpleasant personality traits that were not immediately apparent may surface.

To lessen the chance of misinterpreting the personality of potential partners who may be the unduly aggressive type, it can be revealing to observe their behavior in social and professional situations outside the scope of the proposed partnership. If it differs significantly from their approach to dealing with the partnership, be careful.

Assertiveness

Assertive people are less self-centered than aggressive types. They are primarily focused on their own, personal objectives, but realize that others can contribute greatly to accomplishing these objectives.

They will not hesitate to firmly state their position, but have an open mind and are genuinely interested in finding ways to accommodate the objectives of other team members. They will try to devise a solution and then persuade others of its desirability, but are willing to compromise if necessary.

Assertive people believe that they are credible and trustworthy, and it is important to them that they are perceived as such by others. They are reasonably good listeners because they realize that they need to understand the other team members' points of view in order to persuasively attempt to modify it to achieve their mutual objectives. If they see something that in their opinion needs to be changed, assertive people will take steps to change things and convince other team members to join them in doing so.

Assertive people are good candidates for aircraft partnerships because they tend to have a clear vision of what they seek from a partnership and an equally clearly understanding of what the other partners need out of it to make it work. If there are problems they will stick to the facts and will work to resolve them. If a partnership doesn't work out for them they will tend to take steps to move on.

Reactiveness

Reactive people are extremely accommodating to others, but they also have a clear idea of their own objectives. They actively seek an understanding of the needs and aspirations of others and then see if their own objectives are compatible with the larger interests of the team. They tend to make an effort to find an acceptable way to modify their own position to resolve differences before attempting to convert others to their approach. They recognize the strengths of others, respect alternative opinions, and endeavor to make full use of them to the benefit of the team.

An airplane partnership of reactive people is possibly the most affable and smoothly functioning partnership to be had. Reactive people will not argue passionately to convert other potential partners to their view of what the partnership should be. If they feel that what everyone else wants is acceptable to them they'll sign up. If they have issues with the partnership to be formed they will make a strong effort to accommodate the others, but they won't give away the store. If they feel that their attempt to accommodate isn't successful they won't hesitate to pass on the partnership.

Reactive partners often get along well with assertive types because the two personalities can be complimentary. The assertive partners take the initiative on tackling any issues that arise. The reactive

Fig. 2-2. *This MIG-15LITT is a partnership airplane, believe it or not.*

partners do not find this threatening; after an honest give and take the mutually most acceptable solution is found. In a long-term relationship, assertive and reactive types often learn from each other's behavior and assume each other's behavior characteristics to some degree. They gravitate toward a hybrid *assertive-responsive* stance to the benefit of the entire partnership.

Nonassertiveness

Nonassertive people are the polar opposite of aggressive ones. They totally subordinate their opinions to others and they seek to avoid responsibility. Aggressive types call them wimps. Nonassertive people are open to being taken advantage of by others.

A nonassertive person can actually be a solid member of an airplane partnership, but it is fair to accept them into one only if the other partners have the moral fiber to treat them justly and refrain from taking advantage of them. Nonassertive types should make every effort to realize their nonassertiveness, and if they can't modify their behavior they may be better off staying away from an airplane partnership.

Goals and Expectations

Having become aware of the overriding importance of human compatibility in putting together aircraft partnerships, and gaining some idea of what to look for in personality traits, you can assemble a potential group of partners and begin to hash out what each of you expects from a partnership. Setting initial goals and outlining expectations serves two purposes. It clarifies what the partnership will ultimately be and it gives the potential partners an opportunity to assess how well they are likely to work together.

As you go through the process you and your potential partners should each independently determine your goals and expectations. It is a good idea to do it in writing and to rank each issue in terms of importance to you, say on a scale of 1 to 5; 1 being not important, 5 being very important. Then get together with your potential partners and begin the collective planning process by comparing notes point by point.

In this initial exercise you and your potential partners will begin to clarify the partnership's goals and lay out a framework of analysis. The tools provided in this book will enable you to complete the analysis. When you've gone through the book, completed planning the partnership, and are ready to create, go through the issues again and rate them independently in terms of importance. Compare each other's results and check them against the results of the first pass you made through the issues when you first started discussing the idea of a partnership.

The issues that need to be addressed to develop a credible partnership plan include

- flying objectives
- the airplane
- finances
- partnership structure
- insurance
- operations and safety
- maintenance
- the human factor

Flying objectives

Mission statement   What do you want out of flying? Write a single sentence that summarizes your expectations. "Weekend VFR flying within a 200 mile radius of home base," "Recreational aerobatics," and "Business flights in the Northeast two or three times a week, and one or two long trips a year with my family" are three examples. Each requires substantially different aircraft.

The airplane

Aircraft type   Select the aircraft that best fits the use to which you intend to put it. If you underbuy, you'll be reluctant to do the flights you had in mind because of incompatible aircraft characteristics. Planning to regularly fly IFR without an autopilot in a 118-hp airplane is not going to happen. If you overbuy you'll probably end up being reluctant to put the airplane to a use that is trivial compared to its capabilities. If you stretched financially to buy the airplane you may not have sufficient funds to use it as much as you would like to.

Avionics and equipment   What level of avionics and equipment do you want? What level do you need? Buy what you need. Also, never buy an airplane for the avionics it has in it. Buy an airplane that is in good mechanical condition and add the extra avionics you need.

Future equipment additions   Will you want to add substantial avionics to the airplane in the foreseeable future? This is a perfectly good plan, but your partners must know and agree.

Any special requirements of the aircraft   Do you or any other partners want to use the airplane for something unusual, such as towing gliders, air-to-air photography, or volunteering to ferry ill patients for a regional Angel Flight? Your partners should know up front.

Finances

Amount available for investment   How much money do you have to buy part of an airplane? This is the cash you have from your own resources to hand over to the seller. The costs of storing, maintaining, and operating the airplane are separate.

Financing need   Do you plan to borrow any money to buy the airplane? How much? Are your partners also planning to borrow?

Bank financing option   If you plan to borrow, will you borrow from a bank? Be aware that the bank will want the airplane as collateral and all partners have to co-sign the loan. Each of you will be on the hook for all the others. Your partners may not find this acceptable.

Partner financing option   Are you willing to borrow from another partner or lend to one? Depending on the partners' attitudes this may not be a good idea. It creates inequality in the partnership.

Fig. 2-3. *This rare aerobotic Bonanza is ideal for partners who need a traveling machine but also enjoy recreational aerobotics.*

Financing independent of the partnership Are you planning to get financing for the airplane from sources that have nothing to do with the airplane, such as a home equity loan? This option will not require the airplane to be pledged as collateral and will not make other partners responsible for your debt. It may be the best financing option.

Annual available flying budget What is the amount of money you have available for annual flying expenses? Is it likely to be sufficient to meet your flying objectives? (You'll find out in the next chapter, "Dollars and Sense.")

Partnership structure

Number of partners desired How many partners would you ideally like to have? Why? If your budget is tight, you'll have to work through the numbers to see how many you need.

Share of the aircraft Will all the partners own equal shares in the airplane (recommended)? Occasionally some partners may be willing to buy a larger share in an airplane. In principle this does not sounds like much of a problem (they all get the airplane they want), but it creates inequality in the partnership that could lead to friction.

Under special circumstances owning unequal shares of the airplane may work well, such as when the minority partner has highly demanded skills to contribute to the partnership, such as an airframe and power plant (A&P) certificate or an instructor's rating. Family members who get along well have also formed successful partnerships where one partner has a minority share (there are also plenty of family partnership horror stories).

Legal form of the partnership   Should the partnership be unincorporated or incorporated? Should it have a formal partnership agreement?

Expense sharing mechanism   How will the expenses be shared This is an important area to focus on, because while it is a fairly simple issue, thinking on this issue tends to be a bit fuzzy.

Dispute resolution   How do you envision the resolution of disputes? By consensus or some sort of majority? There are arguments for both, depending on the dispute.

Partnership meetings   How frequently and formally do you envision holding partnership meetings? There are any number of solutions, but all the partners must be agreeable to the option chosen.

Insurance

Level of coverage   How much insurance coverage are you planning to get? Hull insurance, the replacement cost of the airplane, is usually easy to agree upon. Agreeing on the level of liability coverage may be a bigger challenge. The difficulty is usually caused by different levels of wealth among the partners that need protection. "Buy the most insurance you can afford" is a common piece of advice. Doing that and sharing the liability portion proportionally to the level of protection sought by the different partners may be the solution.

Operations and safety

Aircraft use   You need to develop a sense for the likely use patterns of the aircraft. When will the peak times be, what duration will the flights be, will there be a lot of weekend or weekday overnight use? This information will give you a handle on how elaborate your scheduling system needs to be.

Comfort with proposed use   All the partners should be comfortable with each partner's proposed use of the airplane. Generally this is not a big problem, but an issue can arise if one partner doesn't want the airplane used for training and another envisions teaching his three teenagers to fly in it.

Aircraft scheduling   Based on the anticipated use pattern, how do you envision scheduling the aircraft? The weekend is usually the period in most partnerships that needs careful planning. The ideal situation is to have partners with scheduling needs complementary to yours, such as midweek versus weekend use.

Fueling   One of the operational issues that can turn into a mess is the arrangements for fueling the aircraft. Shared fuel accounts can become a real puzzle to sort out. Keeping fuel accounts separate and requiring the airplane to be left fueled after each flight is the best solution, but even then confusion can sometimes occur when you all keep your individual fuel accounts with one FBO.

Expense tracking mechanism   You need to figure out how expenses will be tracked and billed, and how payments will be collected. Many partnerships don't pay enough attention to this requirement and end up with infrequent large payments by partners, which, like accumulated credit card debt, can become problematic.

Assigning tasks to partners   Partners bring different talents to a partnership. Some may be skilled in keeping financial records and others may be good at looking after maintenance. Usually the various tasks in a partnership gravitate to the partner best equipped to handle them. However, it is better to take the initiative and specifically address the issue.

Flying by nonpartners   Are you willing to allow nonpartners to fly the airplane as pilot in command? This can be a touchy issue because all the partners may end up being liable for any mishap caused by a nonpartner who was allowed to fly by one of the partners. On the other hand, if there is appropriate insurance coverage and the specific circumstances warrant it, it may make perfect sense to let a nonpartner fly. But all the partners need to be in agreement.

Maintenance

Maintenance shop preferences   There are big cost differences between having your airplane maintained by a big flashy FBO that caters to airplanes all the way up to corporate jets and the FBO at some small out of the way strip specializing in light aircraft. Plenty of light aircraft are maintained by flashy FBOs, where the annual inspection and related maintenance on a Cessna 150 can end up costing one-third of the airplane's value (a true story). Excellent maintenance is available at the more modest FBOs if you do your homework and choose the right one. At any rate, either choice is fine, but all the partners must think alike.

Owner assistance of shop maintenance   Some shops allow owners to assist in maintenance. A lot of the donkey work can be done by noncertified personnel under the supervision of a certified mechanic (this is the essence of the apprentice system). Owners can save a lot on labor costs working with a shop that allows them to participate in such tasks as removing the dozens of inspection panels for an annual check. You need to decide as a partnership if you want to go that route, and if every partner is expected to participate (some may not have the time or the interest).

Owner maintenance under FAR part 43   Will you perform any preventative maintenance authorized to be done by the holder of a pilot's certificate under FAR part 43? Some pilots are keen to do as much maintenance themselves as possible. Others feel that it should all be left to maintenance professionals.

Discretionary maintenance   This is an important point to consider because it makes a big difference in maintenance costs. There is a significant degree of discretion in how and when certain maintenance items get done. Do we stop-drill the crack in the wingtip fairing and replace it later if the crack gets worse, or do we replace it now? Do we repair the cracked carburetor heat box, or do we get a new one? Do we get a brand new starter or a reconditioned one? The choices and cost differences are many; the partners need to be on the same wavelength.

Cosmetic care arrangements   Are you planning to have the partners wash and wax the airplane periodically, will you hire someone to do it, or will you just let it sit? How often and how formally do you want to schedule cleaning parties? Is a new paint job and a new interior in the cards, and if so, when and at what cost (different options carry different price tags)? All these issues require general agreement among the partners.

Human factors

Available time commitment   Do you have a lot of time to devote to the partnership? Do you want to spend a lot of time at the airport fussing over the airplane? This is both a time and a personality issue. If you have less time to spend taking care of the airplane than your partners expected, you could have a problem. If you are the kind of person who constantly finds something on the airplane to fuss over instead of flying it, and your partners are not, they may become annoyed with you.

Coveting additional equipment Are you a gadget freak who constantly wants to add equipment to the airplane? Are you one of those people who is always on the lookout for the newest gizmo with which to adorn the panel at great cost? Your partners may be happy with the way the airplane is equipped, and they may not be in a financial position to fork over a check every time you visit the pilot shop. If you are a gadget freak, be sure to find like-minded partners.

Importance of aircraft aesthetics Are the aesthetics of the airplane important to you? This is an important personality question. If you want to rush out and get a paint job or an interior because the old one is beginning to fade, you will not be happy with partners who take good mechanical care of the airplane but care little about aesthetic appearance.

Detail orientation Are you a detail oriented, letter of the law fussbudget, or do you just want to get the job done? Within the letter of the law there are a variety of alternatives in flying the airplane and taking care of it. Flying is a schizophrenic affair. It demands engineering precision from people with an artistic bent. To fly a 747 by the numbers is engineering. To fly a Stearman well is art. To some degree all of us who fly are both engineers and artists. But too much of one personality can aggravate too much of the other. If you are intolerant of equally accepted alternative in-flight procedures to the ones you prefer, if you insist that the straps be just so when you leave the airplane, or if you are sensitive about the knots you use to tie down the airplane, be sure to get like-minded partners.

Scheduling structure preference Are you laid back about scheduling, or do you want a highly structured scheduling system? Depending on the use of the airplane and the number of partners, you may be able to get by with informal phone calls to schedule flying time or you may need a rigid booking system. Be sure to understand your needs and your preferences. Some pilots like a formal scheduling system even when an informal arrangement would do. Others prefer to muddle through even when they should have a well-organized reservation systems.

Partner skills brought to the partnership What skills do you and your partners bring to the partnership? This is a very important question. Partners with applicable skills can save a partnership a bundle. Perhaps the ideal partnership is between an A&P mechanic, a lawyer, an accountant, an avionics technician, and a certified flight instructor. You get the idea. Be on the lookout for potential partners who are mechanically inclined or are good with numbers and accounts, or have ratings from which other partners can benefit.

Relative experience level of the partners Would you feel comfortable with partners who are considerably more experienced pilots than you are? Less experienced partners can learn a lot from partners with more experience, but depending on the respective personalities, may end up feeling stifled and ill at ease instead. Pay particular attention to personality issues if there is big difference in experience levels among partners.

Relative wealth of the partners Would you feel comfortable with partners who are much better off financially than you are? Partners who are well to do and are quick to spend may make you feel uncomfortable if you are in a different financial league, even if they have the best intentions.

Socializing Do you want to socialize with your partners, or do you want the partnership primarily for financial convenience? It is worthwhile for all partners to have similar attitudes about socializing with one another, otherwise it is inevitable that someone will be the odd person out.

Goal and value compatibility Are your potential partners' answers to these questions in line with yours? Do they seem to share the values you think are significant? This perhaps is the most important question of them all.

Conflict resolution basics

In the end, no matter how compatible you all are and how smoothly your partnership functions, some conflict is bound to arise. It is important to realize that this is perfectly natural, an integral part of interaction among human beings, to be handled as routinely as any other issue. If you accept that premise, you are well on your way to managing conflict effectively. Below are a few basic observations and guidelines to understanding and resolving conflict within your airplane partnership.

Conflict is a violation of expectations Conflict arises when something happens differently from the way it was expected to occur.

When a person's or group's expectations are not met and the differences can't be cleared up quickly and informally with a brief discussion, the conflict can become sufficiently serious to need the entire group's cooperative attention.

Conflict occurs over issues   People have a tendency to personalize conflict. "He doesn't like my suggestions because he doesn't like me." This is wrongheaded and destructive. He doesn't like the suggestions because in some form or fashion in his opinion they don't offer the solutions for the issues he sees. He may not be able to articulate clearly the specific inadequacies as he sees them or you may not be listening carefully enough to fully comprehend his objections. Focus on the issues!

Conflict is a group problem   When a conflict arises it reduces the entire group's ability to function effectively. It therefore requires the entire group's participation and support to resolve. No individual member should feel as if he or she is the one to have to single-handedly solve the conflict.

Assigning blame is not the point   It doesn't matter whose fault the conflict is. No conflict is resolved by looking for someone to blame. Concentrate on finding a solution, not assigning blame.

Finding alternatives is the point   Conflict is a violation of expectations. The issues that caused the conflict have interfered with meeting expectations. Therefore alternatives must be found that are effective in meeting expectations. Finding acceptable alternatives is at the heart of conflict resolution.

All opinions deserve equal consideration   In the search for alternatives, every member's opinion deserves equal consideration and debate. All opinions should be considered on their merit, on the issues.

Complete openness is most productive   The most productive way to resolve a conflict is through complete openness in a group meeting or series of meetings. All opinions should be aired in an atmosphere of respect and honest debate. All parties should work hard to avoid being defensive. Defensiveness tends to personalize the issues and prevents arriving at a solution.

Compromise is essential   The ability to compromise on a solution that best satisfies the objectives of all parties is essential to resolving conflict. All parties should have a clear understanding of what issues are the most important to them and where they can give.

The role of a mediator   If passions run particularly high it may be useful for a member of the group to act as a mediator who carefully clarifies each party's opinions and alternative proposals and manages the debate.

Document the resolution   When a solution to the conflict is reached it is important to clearly document it to accurately reflect the alternative accepted by the group so that there are no misunderstandings. Formal amendments of the partnership agreement or any written operating rules may also be necessary.

3

Dollars and Sense

The most compelling reason to enter into an aircraft partnership is financial, and a clear understanding of the costs of flying is critical to devising the best partnership solution. Under the right circumstances an aircraft partnership will greatly reduce the cost of owning and flying an airplane in comparison to owning one on your own or renting one. The key term is "the right circumstances." The fact is, if you don't fly a sufficient number of hours, even a partnership is going to be more expensive than renting an airplane at your friendly FBO. If, however, a severely limited number of hours in a rental airplane with all the attendant hassles and restrictions is not appealing enough to give you what you expect from flying, an airplane partnership may well mean the difference between continuing to fly or grounding yourself.

The cost of flying in a partnership has to be compared to two alternatives: the cost of owning an airplane alone and the cost of renting one. Table 3.1 is a graphic illustration of the relationship between the per pilot costs of renting or owning an airplane in partnerships of up to six people for a given number of hours flown. Some trends are immediately noticeable:

- Depending on the number of hours flown and the number of partners, there comes a point when it is less expensive to own than to rent.
- The more partners there are the less the per hour cost of flying per partner.
- As the number of partners increases, the rate at which the cost per partner decreases diminishes. This suggests that there may be a point where the financial gain of adding more partners may

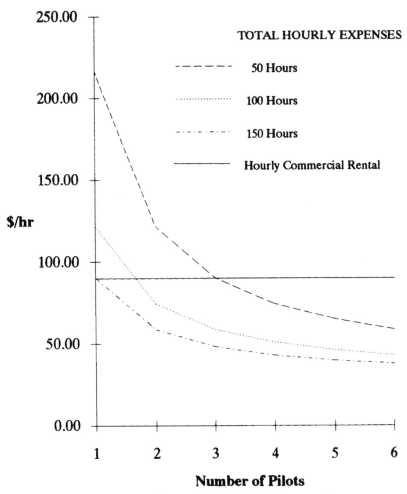

Table 3.1. *Comparison of aircraft rental and partnership costs.*

be outweighed by the nonfinancial drawbacks of operating with a large number of partners.

To appreciate how dramatically a partnership can reduce the cost of flying relative to both alternatives look at Table 3.2. It shows in more detail than the graph the annual cost per pilot of flying a 1976 Piper Archer for 100 hours as a member of partnerships of two to six pilots, compared to flying the same airplane as a single owner or renting it at $85 per hour from an FBO (the costs were determined by using the financial model for calculating the cost of flying which is presented later in this chapter).

You can see that with two partners, the per partner cost of flying 100 hours a year is $7165 compared to the $11,110 it would cost

as a single owner and the $8500 it would cost per pilot for an equiv-
alent amount of rental time in the same airplane. This is a savings of
$3945 per year over the single ownership option and a savings
of $1335 per year over the rental option. The savings over single
ownership is a strong incentive to find ways to meet the nonfinancial challenges of making a partnership work. And compared to the
hassles and restrictions of renting, the advantages of owning in a
partnership are particularly attractive.

There is another very important way in which an airplane partnership can benefit many of us. It can vastly improve the bang for
our buck by allowing us to afford a higher quality of flying than what
we could manage on our own. For the same amount of money that
it would cost you to own and fly a Piper Archer alone, three of you
could fly a Bonanza of similar vintage in a partnership. Table 3.3
shows the cost of flying 100 hours per year per pilot as a single
owner of the Archer and as a member of a three-way Bonanza partnership.

These figures should begin to convince even the financially wool-
liest aviator among us of the benefits of aircraft partnerships. They
should also begin to get across the idea that thoroughly understanding the costs of flying is essential to accurately determine the best
possible flying alternatives given your objectives and resources. The
next section presents the ins and outs of figuring the dollars and cents
of aircraft ownership and operation.

1976 Piper Archer PA-181	Per Pilot Annual Cost of Flying 100 Hours ($)					
Number of Pilots:	1	2	3	4	5	6
Partnership Cost Per Pilot	11110.00	7165.00	5850.00	5192.50	4798.00	4535.00
Rental Alternative Per Pilot	8500.00	8500.00	8500.00	8500.00	8500.00	8500.00

Table 3.2. *Per pilot cost of flying 100 hours in partnerships of up to six pilots.*

Aircraft Make and Model:	1976 Piper Archer	1976 F-33 Bonanza
Number of Pilots:	1	3
Per Pilot Cost of 100 Hours	11110.00	11077.00

Table 3.3. *Three partners can fly a Bonanza for the price of flying an Archer alone.*

Flying Cost Basics

Look at Table 3.4. It portrays the financial structure of owning an airplane with up to six partners. The airplane chosen for the example is

1976 Piper Archer PA-181   **AIRCRAFT PARTNERSHIP FINANCIAL ANALYSIS ($ per pilot)**

NUMBER OF PILOTS	1	2	3	4	5	6
CAPITAL INVESTMENT	**50000.00**	**25000.00**	**16666.67**	**12500.00**	**10000.00**	**8333.33**
LOAN AMOUNT	**0.00**	**0.00**	**0.00**	**0.00**	**0.00**	**0.00**
ANNUAL FIXED EXPENSES						
Tiedown/Hangar	1020.00	510.00	340.00	255.00	204.00	170.00
Insurance	1500.00	750.00	500.00	375.00	300.00	250.00
State Fees	120.00	60.00	40.00	30.00	24.00	20.00
Annual	750.00	375.00	250.00	187.50	150.00	125.00
Maintenance	1000.00	500.00	333.33	250.00	200.00	166.67
Loan Payments	0.00	0.00	0.00	0.00	0.00	0.00
Cost of Capital (non-cash)	3500.00	1750.00	1166.67	875.00	700.00	583.33
Total Fixed Expenses / yr	**7890.00**	**3945.00**	**2630.00**	**1972.50**	**1578.00**	**1315.00**
HOURLY OPERATING EXPENSES						
Fuel	16.80	16.80	16.80	16.80	16.80	16.80
Oil	0.40	0.40	0.40	0.40	0.40	0.40
Engine Reserve	10.00	10.00	10.00	10.00	10.00	10.00
General Maint Res	5.00	5.00	5.00	5.00	5.00	5.00
Total Op Exp / hr	**32.20**	**32.20**	**32.20**	**32.20**	**32.20**	**32.20**
TOTAL EXPENSES PER HOUR						
50 Hours flown per year	190.00	111.10	84.80	71.65	63.76	58.50
100 Hours flown per year	111.10	71.65	58.50	51.93	47.98	45.35
150 Hours flown per year	84.80	58.50	49.73	45.35	42.72	40.97
Hourly Commercial Rental	**85.00**	**85.00**	**85.00**	**85.00**	**85.00**	**85.00**
TOTAL ANNUAL EXPENSES						
50 Hours, Own	**9500.00**	**5555.00**	**4240.00**	**3582.50**	**3188.00**	**2925.00**
Own (cash only)	6000.00	3805.00	3073.33	2707.50	2488.00	2341.67
Rent	4250.00	4250.00	4250.00	4250.00	4250.00	4250.00
100 Hours, Own	**11110.00**	**7165.00**	**5850.00**	**5192.50**	**4798.00**	**4535.00**
Own (cash only)	7610.00	5415.00	4683.33	4317.50	4098.00	3951.67
Rent	8500.00	8500.00	8500.00	8500.00	8500.00	8500.00
150 Hours, Own	**12720.00**	**8775.00**	**7460.00**	**6802.50**	**6408.00**	**6145.00**
Own (cash only)	9220.00	7025.00	6293.33	5927.50	5708.00	5561.67
Rent	12750.00	12750.00	12750.00	12750.00	12750.00	12750.00

ASSUMPTIONS:				
	Hangar/Month:	85.00	Fuel Cost ($/gal):	2.10
	Insurance/yr:	1500.00	Fuel Cons (gal/hr):	8.00
	State Fees/yr:	120.00	Oil Cons (qt/hr):	0.20
	Annual:	750.00	Oil Cost ($/qt):	2.00
	Maintenance/yr:	1000.00	Engine MOH Cost:	15000.00
	Loan Amount:	0.00	Time Rem. To OH:	1500.00
	Loan/Inv Term-yrs:	10	Gen Maint Res/hr:	5.00
	Loan Interest Rate:	12.00%	**Aircraft Value:**	**50000.00**
	Cost of Capital/yr:	7.00%	**Com. Rental/hr:**	**85.00**

Table 3.4. Aircraft partnership financial analysis.

a 1976 Piper Archer II in good condition with 500 hours on the engine since major overhaul. At the time of writing the Archer was valued at $50,000.

A word of encouragement: If you are not the financial type, don't be intimidated by financial analysis. The only skills you need to calculate the cost of flying are the ability to add, subtract, multiply, and divide. If you can do the basic math required to navigate and track fuel consumption, you have the skills to figure out the dollars and cents. And a word of caution: Don't get too hung up on the specific assumptions in the example. They were valid at one point in time in one location. Concentrate instead on the method of analysis to see how you can put it to work on assumptions valid for your own situation.

The first item you'll notice is the capital requirement. A lack of capital will keep you from buying an airplane, but the greater the number of pilots in the partnership, the less capital you'll need. In this example you can buy a one-third share in an Archer for the cost of a modest car. And, just as you can when you buy a car, if you don't have enough capital you can take out a loan to finance part of the airplane. You will find, however, that borrowing can be quite expensive. In this example we chose to be conservative and assumed that no loan was taken out to buy the airplane.

Beyond the capital investment, the basic cost components of aircraft ownership are *fixed expenses* and *operating expenses* (I'll define them in a moment). By developing reliable information on these two cost components you can accurately determine per pilot the *total expenses per hour* of flying an airplane as well as the *total expenses per year* (total annual expenses) spent on flying. To calculate these costs you need to gather some basic information, which is presented in the assumptions section of the table.

Before we delve into the details of financial analysis let's define the main components you'll have to calculate:

- *Fixed expenses* are expenses that you have to pay for when owning an airplane regardless of whether or not it flies. Good examples are hangar costs and state fees. In the model I measure these expenses annually. The conventional way of looking at them.

- *Operating expenses* are expenses that you have to pay only as a direct consequence of flying the airplane, such as for fuel and oil. I measure these expenses per hour in the model, because you keep track of your flying on an hourly basis. To get operating expenses for the year, multiply the hourly operating expenses by the number of hours flown in the year.

Fig. 3-1. *Cherokee 140 and Cessna 150, two recreational favorites.*

- *Total expenses per hour* is your cost of flying per hour, given the number of hours flown during the year. This figure is most commonly used to compare the cost of airplane ownership to the alternative of renting an airplane. To get total expenses per hour do the following calculations: divide the annual fixed expenses (total fixed expenses per year) by the number of hours flown per year. This gives you your fixed costs per hour. The more you fly the less the fixed costs per hour. Add to this amount to the hourly operating expenses (total operating expenses per hour) for your total hourly expenses.

- *Total annual expenses* is the bottom line. It is the amount it costs you to fly for a year, taking into account both the fixed expenses and the operating expenses for the year. Fixed expenses are derived on an annual basis. To get total annual expenses take total expenses per hour (see above) and multiply this figure by the hours flown per year.

Once you get through this chapter you can make copies of the blank worksheet in appendix 2, collect your own assumptions, and calculate the costs relevant to your options. A highly recommended alternative is to build your own financial spreadsheet that will do the calculations for you. The model is presented in appendix 2. It contains all the formulas for building the spreadsheet in Microsoft Excel. Once you've built the model, all you need to do is key in the assumptions and the spreadsheet will calculate all the values for you. It is well worth building the spreadsheet because it vastly increases your ability to analyze alternatives. As you change any assumption the associated values are instantly recalculated.

Now let's take a detailed look at the various cost components, the issues they raise, and the savings opportunities they present for a partnership depending on how they are handled.

Fixed Expenses

The fixed expenses, the ones incurred regardless of whether or not the airplane flies, are relatively straightforward to assess, with a few exceptions. Note that the share per partner of fixed expenses is the single-owner expense divided by the number of partners (see Table 3.5). The majority of partnerships follow this equal allocation of fixed expenses per partner (later I will present alternative ways of sharing some of these expenses which are practiced by a small minority of partnerships where there is a great disparity between the amount of time each partner flies or the partners' experience level).

NUMBER OF PILOTS	1	2	3	4	5	6
ANNUAL FIXED EXPENSES						
Tiedown/Hangar	1020.00	510.00	340.00	255.00	204.00	170.00
Insurance	1500.00	750.00	500.00	375.00	300.00	250.00
State Fees	120.00	60.00	40.00	30.00	24.00	20.00
Annual	750.00	375.00	250.00	187.50	150.00	125.00
Maintenance	1000.00	500.00	333.33	250.00	200.00	166.67
Loan Payments	0.00	0.00	0.00	0.00	0.00	0.00
Cost of Capital (non-cash)	3500.00	1750.00	1166.67	875.00	700.00	583.33
Total Fixed Expenses / yr	**7890.00**	**3945.00**	**2630.00**	**1972.50**	**1578.00**	**1315.00**

Table 3.5. *Annual fixed expenses per pilot.*

Tie-down or hangar

This is what you pay for storing the airplane. Most airports bill you monthly. Multiply the monthly bill by 12 for the annual figure. In this example the Archer is tied down outside for $85 per month at a major airport with good services and easy access for the partners. Note how effective the partnership is in reducing storage costs per partner. With three other partners you each pay only $255 for the whole year; that is a mere $21.25 per month!

Many partnerships find that sharing the storage fee results in such a cost reduction per person compared to single ownership that it becomes affordable to hangar the airplane. Take advantage of this opportunity if it is affordable, because it is one of the best ways to protect your investment.

Particularly budget-minded partnerships, on the other hand, may opt for a tie-down at a less expensive airport to cut storage costs to the bone, even if it is less convenient. A $40 per month tie-down cuts each partner's share in a four-way partnership to $10 per month. Some pilots may say that saving $10 per month is not worth the extra inconvenience. Others may say that $10 per month per partner is $120 per year, and $120 per year times four partners is $480, a tidy sum for to add some goody to the airplane once a year. The permutations are many; think creatively to design the structure that is best for you.

Insurance

This is your annual insurance bill. The amount varies considerably, depending not only on equipment, but also on level of coverage, the experience level of the partners, use of the airplane, insurer, and more. It is said that you get what you pay for, and in our litigious society attempting to save on insurance can be penny-wise and pound-foolish. Nevertheless, call several brokers for competing quotes. Depending on the state of the insurance market the quotes may vary significantly for generally the same level of coverage. In our example a policy with a low deductible, high per person per event limits, and a $1 million total limit is $1500.

State fees

Some states charge periodic recurrent fees and taxes. This is a catch-all category for such expenses, which will vary greatly from state to state. Annual registration fees and excise taxes fall in this category. Call your state aeronautics commission for more information. If you don't call them, they'll call you. Some state commissions live to scrutinize new FAA registrations and swoop down on the unwary with bills that can cause annoying surprises. Some even send out covert raiding parties to roam the airports in search of local aircraft without state fee stickers.

Annual inspection

This expense is the cost of the annual inspection only. This is what your friendly mechanic charges you just to look at the airplane and determine if any maintenance work needs to be done. It is a fixed fee, set by the size and complexity of the airplane, and is listed separately on your annual bill. Any work that is done as a result of the inspection is extra and is accounted for in *maintenance* and/or under hourly operating expenses in General Maintenance Reserves. The annual inspection fee for the Archer in the example is $750. This figure will vary somewhat from region to region and mechanic to mechanic. For planning purposes ask your mechanic what the fees are for the airplanes you are considering.

Maintenance

This expense category is the most difficult to estimate. Many pilots will argue that it is not even a fixed expense because maintenance is a function of operating the airplane. They are technically correct, and in the case of airplanes that fly a lot (such as those in commercial or corporate use), a per hour maintenance reserve paid in under hourly operating expenses can be quite accurately estimated to meet maintenance expense.

However, an airplane in a noncommercial aircraft partnership may not fly enough to enable the partners to accurately predict an adequate amount of maintenance reserve that needs to be set aside per flying hour. For such partnerships it is best to set aside an annual fixed sum as a maintenance reserve in addition to paying in a per hour maintenance reserve. The fixed annual payment into the maintenance reserve provides a cushion for surprises. The maintenance reserve amount paid in per flying hour (under operating expenses) can be designated to take care of such normal wear and tear items as brake pads, tires, spark plugs, oil and fuel filters, the replacement of deteriorating baffling, lightbulbs, and the like.

It is perfectly acceptable to consider all maintenance as an operating expense and reserve for it per flying hour. But whichever route you choose, be sure to reserve enough. Most partnerships tend to underreserve for maintenance and then need to scramble around with an assessment when a cylinder blows, a gyro fails, or a radio refuses to transmit.

Fig. 3-2. *A brand new Turbo Saratoga can be affordable for many pilots in a partnership.*

In this example I've set aside a maintenance reserve of $1000 as a fixed annual expense for the Archer. This amount is in addition to the $5 per hour maintenance reserve that you will see under operating expenses, and is available to pay for anything that is uncovered by the annual inspection or during the course of flying.

Loan payments

This expense is the payment you make in a year on any loans you've taken out to buy the airplane. Most loan payments are monthly, so multiply the monthly sum by 12 to get the annual payment. The Excel spreadsheet presented in appendix 2 has the payment formula built into it. You need only to enter the loan amount, the number of years, and the interest rate, and the spreadsheet will calculate and annualize the payment.

Alternatively, if you haven't built the spreadsheet and would like to do projections, you should call a banker who makes aircraft loans and get a quote for the payment amount based on how much you want to borrow for how long at what interest rate.

Cost of capital (noncash)

The cost of capital can be confusing to pilots unfamiliar with finance, but it is a simple concept. It is the income you are foregoing by investing your money in your airplane instead of putting it into an investment that earns a higher return. A typical example would be to take the $25,000 put in the Archer for a half share and investing it in a mutual fund that earns 10%. Had you put the money into the mutual fund, you would have had $27,500 in your account by the end of the year instead of half an Archer at the airport.

Assigning a value to the cost of capital is at best a guesstimate, but you shouldn't kid yourself that you aren't foregoing some income by putting the money in an airplane. It is true that the airplane itself is an investment that may also appreciate in value, but this appreciation must be compared to forgone appreciation from alternative investments into which you would have most likely put your money otherwise. In this example I have done just that. I have said that the Archer would appreciate 3% in a year, while a solid mutual fund would have increased by 10%. Thus the cost of capital in our example is 10% minus 3%, which is equal to 7%. Note that this is a *non-cash* item because it is not a cash expense that you have to pay out of your pocket. Rather it is money that will not flow into your pocket. Therefore the cash that you will actually have to come up with for your flying expenses for a year excludes this amount from the total cost. If you are only interested in how much cash it is going to cost you to fly per year you may choose to ignore this item.

Total fixed expenses per year

This expense is simply the total amount of fixed expenses per year. These expenses have to be spent regardless of how much or how little the airplane flies. Note that fixed expenses per partner decline rapidly as the number of partners increases because these are expenses shared by all partners.

Operating Expenses

Look at Table 3.6. It is a list of operating expenses. These expenses are incurred only when the airplane flies, and then only by the partner flying the airplane. Unlike fixed expenses, which are shared by the partners, these expenses are *not* shared. Rather, by paying them in full, each partner is paying for his or her use of the shared airplane. These expenses are not affected by the number of partners or the hours flown.

Fuel

This is the amount you have to spend on fuel for every hour of flight time. It is calculated by taking the airplane's fuel consumption

NUMBER OF PILOTS	1	2	3	4	5	6
HOURLY OPERATING EXPENSES						
Fuel	16.80	16.80	16.80	16.80	16.80	16.80
Oil	0.40	0.40	0.40	0.40	0.40	0.40
Engine Reserve	10.00	10.00	10.00	10.00	10.00	10.00
General Maint Res	5.00	5.00	5.00	5.00	5.00	5.00
Total Op Exp / hr	**32.20**	**32.20**	**32.20**	**32.20**	**32.20**	**32.20**

Table 3.6. *Hourly operating expenses per pilot.*

in gallons and multiplying it by the cost of fuel per gallon. The Archer in the example burns about 8 gallons per hour, at a cost of $2.10 per gallon, resulting in a fuel cost of $16.80 per hour.

Partnerships have various schemes whereby each partner pays or accounts for his or her own fuel expenses. The simplest method is for each partner to maintain a separate fuel account at the airport, and require that the airplane be returned to its tie-down or hangar fueled to a certain level.

Oil

This expense is the cost of oil burned by the airplane per hour. It is quite a small sum, but readily identifiable. To calculate oil expenses divide the price of a quart of oil by the amount of oil the airplane uses in an hour.

Engine reserve

The overhaul of an airplane's engine is one of the largest expenses during its lifetime. It is essential to reserve for this eventuality and this is best accomplished by setting aside an adequate engine reserve per flying hour.

This expense is actually one of the easier ones to estimate, provided that you accurately estimate the amount of time remaining on the engine before it needs to be overhauled. The big question is will the engine make it to full recommended TBO (time before overhaul), or will it need to be overhauled earlier. Frequently flown, well-maintained engines traditionally run well beyond the recommended TBO. Engines flown less than 15 to 20 hours per month are less likely to make it to the TBO.

Once you have a reasonable estimate of the engine time remaining and the cost of an overhaul (all the major overhaul shops can quote standard rates; don't forget installation expenses), the required reserve is calculated by dividing the cost of the overhaul by the hours remaining.

In the case of the Archer in our example, the airplane had 500 hours on the engine when purchased by the partnership. It is expected to make it to full TBO, so it has 1500 hours remaining. The cost of overhauling and reinstalling the engine is $15,000. Therefore $10 per hour ($15,000/1500 hours) has to go into the engine reserve to have sufficient funds for the overhaul when the engine runs out.

If the engine would be expected to run out in less time, say 1000 hours, then $15 per hour would have to be reserved to remain on target financially.

Some partnerships that buy an airplane with a midtime engine opt to fund an engine reserve account with money equal to the reserve required to cover the time already used. This option establishes a nice initial cash reserve for the engine and proportionally reduces the amount to be paid in per hour.

If you wanted to do this with the Archer, you would divide $15,000 by 2000 hours, which is equal to $7.50 per hour. You would then multiply $7.50 by 500 (the hours already used), which equals $3750. Each partner would put his or her share of this amount into a common engine reserve account and then would add $7.50 to the reserve account for every additional flight hour. Either of these methods is acceptable; what's important is that all the partners agree on the method.

Some partnerships are easygoing about reserving for the overhaul, figuring that everyone will pay in their share via a special assessment when the time comes. Given the large amounts of money involved, this attitude can lead to considerable financial strain on the partners and friction in the partnership.

General maintenance reserve

The general maintenance reserve is an hourly reserve for routine maintenance items that need to be performed during the year. It requires an estimation of the cost of likely maintenance per year, which is divided into the number of hours likely to be flown during the same period.

As mentioned earlier, some partnerships prefer to have an annual assessment carried under fixed costs in addition to the hourly maintenance. In this case the hourly reserve amount may be reduced. Other partnerships prefer to reserve the entire maintenance amount for the year in this category, which can result in a hefty hourly maintenance reserve charge. And if the hours flown are underestimated, a special assessment may need to be made to cover any shortfalls.

In the case of the Archer in our example, $5 is added to the maintenance reserve per hour flown. If the airplane flies 300 hours per year this reserve will amount to $1500 per year, which is in addition to the $1000 reserved under fixed expenses and the $750 set aside for the annual inspection, resulting in a total of $3250 available for maintenance.

The maintenance reserve is the expense category that you will find yourselves fine-tuning the most often, depending on changing circumstances. If an excessively large reserve starts to build up, you may ease off the hourly amount. If you are chronically caught short, it may be time to increase the hourly reserve. If a major expense looms, such as a paint job or interior replacement, you may also choose to increase the hourly reserve in anticipation of the event.

You are encouraged to devise additional hourly expense categories that may suit your specific needs. For example, many partnerships that fly complex airplanes maintain a separate propeller overhaul reserve.

This completes the financial information that you need to gather to assess your costs of flying. Now let's move on to see what it all means in terms of total expenses per hour and total annual expenses spent on flying.

Total Expenses Per Hour

The total expenses per hour that it costs you to fly under various partnership options has to be compared to the expenses of accomplishing the same objectives as a single owner and as a renter. The total expenses per hour per partner (see Table 3.7) are derived as follows:

- Divide the annual fixed expenses by the hours flown to determine these expenses per flight hour.
- Add the above sum to the hourly operating expense figure.

The hourly rental rate at an FBO remains constant regardless of the hours flown. However, your total hourly expenses as a single owner or as a member of a partnership vary based on the number of hours

NUMBER OF PILOTS	1	2	3	4	5	6
Total Fixed Expenses / yr	7890.00	3945.00	2630.00	1972.50	1578.00	1315.00
Total Op Exp / hr	32.20	32.20	32.20	32.20	32.20	32.20
TOTAL EXPENSES PER HOUR						
50 Hours flown per year	190.00	111.10	84.80	71.65	63.76	58.50
100 Hours flown per year	111.10	71.65	58.50	51.93	47.98	45.35
150 Hours flown per year	84.80	58.50	49.73	45.35	42.72	40.97
Hourly Commercial Rental	85.00	85.00	85.00	85.00	85.00	85.00

Table 3.7. *Total expenses per hour per pilot.*

flown. This is because you have to allocate the fixed expenses to the hours flown, and the greater the number of hours flown and the more the number of partners sharing the fixed expenses, the less the fixed expenses per hour per partner.

Knowing the total expenses per hour allows you to answer several important questions regarding flying the Archer in our example:

- How many hours do I have to fly before a partnership option of my preference is less expensive than the rental alternative? If you prefer a two-person partnership in the Archer you'll have to fly about 100 hours before you are ahead of the rental game. At 100 hours your total hourly expenses are $71.65 compared to the hourly rental rate of $85.

- How large a partnership do I need for the given number of hours per year that I plan to fly to beat the rental alternative? If you estimate that you'll fly about 50 hours per year, you'll need at least a three-person partnership to break even in comparison to renting.

- How favorably do partnerships compare to single ownership on the basis of total hourly expenses? Very favorably. If you fly only 50 hours in a two-person partnership you are saving $79 ($190–$111) compared to what your total hourly expense would be as a single owner. As the number of flying hours increases, the advantage diminishes but nevertheless remains substantial. At 150 hours flown, the advantage is $26 in favor of a two-person partnership over the single owner. For partnerships of larger size the savings increase dramatically.

These questions are only a few examples to get you started in asking your own questions relevant to your circumstances. Think of as many alternatives as you can and compare them carefully to get the solution that best meets your needs given your financial resources.

Total Annual Expenses

Total annual expenses is the bottom line (see Table 3.8). This is the amount it will cost you to fly given the alternative you choose. Approaching it from the other side, if you have a set annual budget, the total annual expenses figure tells you which alternatives are available to you.

For single ownership and partnerships the total annual expenses figure is derived by multiplying the hourly operating expenses by the number of hours flown and adding this sum to the annual fixed

NUMBER OF PILOTS		1	2	3	4	5	6
Total Fixed Expenses / yr		7890.00	3945.00	2630.00	1972.50	1578.00	1315.00
Total Op Exp / hr		32.20	32.20	32.20	32.20	32.20	32.20
TOTAL ANNUAL EXPENSES							
50 Hours,	Own	9500.00	5555.00	4240.00	3582.50	3188.00	2925.00
	Own (cash only)	6000.00	3805.00	3073.33	2707.50	2488.00	2341.67
	Rent	4250.00	4250.00	4250.00	4250.00	4250.00	4250.00
100 Hours,	Own	11110.00	7165.00	5850.00	5192.50	4798.00	4535.00
	Own (cash only)	7610.00	5415.00	4683.33	4317.50	4098.00	3951.67
	Rent	8500.00	8500.00	8500.00	8500.00	8500.00	8500.00
150 Hours,	Own	12720.00	8775.00	7460.00	6802.50	6408.00	6145.00
	Own (cash only)	9220.00	7025.00	6293.33	5927.50	5708.00	5561.67
	Rent	12750.00	12750.00	12750.00	12750.00	12750.00	12750.00

Table 3.8. *Total annual expenses per partner.*

expenses. The annual rental expense is the hourly rental rate multiplied by the hours flown.

The *Own (cash only)* line is the total annual expense excluding the cost of capital (the cost of capital is the amount of money per year that you are not receiving because you put your money in the airplane instead of a higher yielding investment). It shows the actual cash you need on hand to meet the annual expenses.

Here are some of the questions you can begin to answer about the Archer in our example once you know total annual expenses:

- I have a given annual flying budget. Am I better off renting or going into a partnership? That's easy enough to answer for a given airplane type and number of partners. See which alternative will buy you more flying time. If you have $2000 per year, no partnership option in Table 3.8 will buy you more flying time than what you can buy as a rental pilot. If you have $4250 you can buy 50 hours of rental time. That is less expensive than 50 hours in a two-person partnership, but more than 50 hours in a partnership of three or more pilots.

- I want to fly twice as many hours as I could in a two-person partnership for the same amount of money. How large a partnership do I need? Fifty hours in a two-person partnership costs $5555. For $5850 you can fly 100 hours in a three-person partnership. For $5192 you can fly 100 hours in a four-person partnership.

- On a cash basis how large a partnership do I need if I plan to fly 50 hours to beat the rental option? If you choose to ignore the cost of capital (a noncash item), you'll be ahead of the rental option even in a two-person partnership if you fly only 50 hours.

- How significant are my savings under the various partnership options compared to single ownership? In every case the savings run into the thousands of dollars per year.

This chapter explored analyzing the cost of a single airplane. In the next chapter, "Partnership Options," these techniques will be put to work to compare a multitude of airplanes, objectives, and partnership structures to show you how to come up with the best solution for your particular needs and goals. But before moving on let's discuss a few alternatives to sharing certain partnership expenses.

Expense Sharing Alternatives

A good basic philosophy for sharing partnership expenses is to share the fixed expenses equally and have each partner pay for the hourly operating expenses incurred when the partner is flying the airplane. This is an equitable distribution for several reasons. The fixed expenses have to be paid regardless of whether or not the airplane flies. It is the price of admission to airplane ownership. All the partners have an equal potential benefit from owning the airplane. How much each of the partners chooses to fly (exercises that benefit) is up to them.

When a partner flies the airplane he is putting wear and tear on the airplane and is incurring expenses (fuel, oil, etc.) for an experience not shared by the other partners. It is therefore fair that the partner pay entirely for these expenses, including wear and tear in the form of the hourly engine and maintenance reserve payments.

In partnerships where one pilot flies considerably more than another (say 100 hours compared to 50 hours), the issue is sometimes raised that the partner flying fewer hours is subsidizing the one flying more. This is not true, because the partner flying more is paying for his share of the greater wear and tear he is putting on the airplane. Suppose that the engine of the Archer in a two-way partnership in our example is overhauled after having 1500 hours put on it by the partnership. One partner has put 1000 hours on it, the other 500. If the $10 per hour engine reserve was paid in every time the airplane flew, the partner flying more will have contributed $10,000 to the overhaul, while the one flying less will have contributed only $5000.

The idea of having the partner who flies more pay more of the fixed expenses is wrongheaded because these expenses are not a function of flying time. If you still have a problem with this concept, consider two airplane owners who are *not* partners but keep their

Fig. 3-3. *The Zenith is a new contender. (Courtesy of Richard Vander Meulen)*

respective airplanes in a shared hangar. It would be silly for one owner to say to the other that he should pay for only one-third of the hangar because his airplane flies only one-third of the time compared to his hangarmate's airplane. The rationale is the same in a partnership. If you own 50% of the airplane you should pay 50% of the storage and other fixed costs.

There are, however, some scenarios under which paying a differential share of a fixed expense may be justified

Insurance

If the partners have vastly different experience levels the least experienced partners may have to pay a premium over what the insurance would cost if only the more experienced partners were insured. In this case it is perfectly legitimate to ask the inexperienced partner to pay the difference, since he is making it more expensive for everyone.

Maintenance

In partnerships which have chosen to require an equally shared annual fixed maintenance payment (under fixed expenses), the partners flying less are subsidizing the partners flying more because the maintenance expense pays for hourly wear and tear. If the difference in hours flown per partner is significant, it may be best to do away with the fixed maintenance expense and reserve for maintenance entirely under the hourly maintenance reserve.

Inactive partner

From time to time a partner may get into a situation where he or she can't fly for an extended period of time but doesn't want to sell out because the circumstance is only temporary. In that case it might be acceptable to all partners if the affected partner goes inactive and does not pay any of the expenses while inactive. In this case the inactive partner essentially becomes a passive investor. The active partners then have a smaller partnership for the duration with higher operating costs and greater access per partner. To avoid potential conflicts it is advisable to allow a partner to go inactive only over the long term (6 months or more) and only in extreme circumstances, such as an illness, a messy divorce, or a temporary reassignment away from home.

Case Study: Cherokee 140–Four Partners

The aircraft

The partnership's aircraft is a 1971 Cherokee 140 (PA 28-140). It was purchased in 1987 at a bargain price of $10,000, requiring only $2500 from each of the four partners. The price was low because at the time an airworthiness directive (AD) proposal was pending that would have required removal of the wings for inspection. The proposal was withdrawn and the partners benefited from a substantial increase in the aircraft's value in comparison to the purchase price.

Partnership background

The four partners in this enduring partnership know each other well and their personalities have proven to be comfortably compatible over the years. Two of the partners are A&P mechanics, one of whom also has his IA (Inspector Aircraft). The airplane is used mostly for local flying and day trips.

Partnership structure

The partnership is incorporated and operates under a standard partnership agreement (technically the by-laws of the corporation). Each partner owns an equal share of stock in the corporation. The primary reason for incorporation is the benefit of the liability protection it offers. Another benefit is that there is no sales tax on the sale of a partner's share in the corporation (this is somewhat offset by a minimum annual corporate tax that has to be paid, even though the corporation has no income). Decisions are made by consensus.

Finances

The partners share fixed expenses equally. The two big items are $760 per year for insurance and $160 per month for hangaring the airplane. In addition to these items the partners are also each required to pay a fixed expense of $25 per month for incidentals, which includes oil.

Hourly operating expenses are the sole responsibility of each partner for the hours flown by that partner. A maintenance reserve charge of $10 per hour is paid into the partnership and each partner pays for his own gas directly to the FBO. Here's how this partnership breaks down its expenses:

ANNUAL FIXED EXPENSES ($)

	Total	Per Partner
Hangar	1,920	480
Insurance	760	190
Annual Maintenance	300	75
Total Fixed Expenses	2,980	745

HOURLY OPERATING EXPENSES ($)

Fuel	17
General Maint Res	10
Total Hourly Op. Exp.	27

TOTAL EXPENSES PER YEAR ($)

50 hours of Flying Time	2,095
Hourly Rate @ 50 hr/yr	42

The partnership has a joint account run by one of the partners into which all the partners make their payments and from which the bills are paid. A balance of $500 is maintained in the account. If any unexpected expenses come up, an assessment is made to pay for them. The partners have confidence in each other's ability to come up with a large lump sum if necessary. They prefer this arrangement because it doesn't require them to park large amounts of money in an idle airplane account. The system has worked well. When an

Case 3-1. *The Cherokee 140 is a great weekend flier.*

unexpected engine overhaul was required no payment problems were encountered.

The maintenance expense is low because the A&P mechanics among the partners volunteer their services to perform most of the maintenance requirements.

Operations

Scheduling for short, local flights is informal. If the airplane is at the airport it can be taken for a local flight. Given the partners' use patterns, this arrangement has been working well since 1987, when the airplane was bought. For scheduling longer flights a weekly pocket planner is kept in the airplane into which planned trips are entered. The pocket planner is also used as the flight log to track per partner use. At the end of each flight the tank is refilled to the tabs and the FBO is paid directly by the pilot. One partner flies the airplane between 50 and 100 hours per year, the others about 50 hours or less.

Maintenance

Given that two of the partners are A&P mechanics, most of the routine maintenance and the annual inspection are performed by the partnership. The annual inspection, however, is signed off by an outside inspector, even though one of the A&P partners is also an IA.

Three years into the partnership, when the engine had 450 hours on it since major overhaul (which was done before the partners bought the airplane), one of the partners experienced an engine failure and managed to dead stick the airplane back to home base. The partners had to do a major overhaul and had to make an assessment for the costs. Because of the mechanics among them they were able to do about one-third of the work themselves and managed to get away with a total cost of $3600.

The partnership experience

Its long existence is the best proof of this partnership's success. All the partners contend that it is a superior alternative to renting, not only from a financial point of view but also because of the flexibility they have in using the airplane.

No partnership is entirely free of differences of opinion. In this partnership some partners have been more keen to upgrade avionics than others, and it isn't always easy to get everyone together for partnership meetings. All the partners maintain, however, that these issues are negligible and are far outweighed by the partnership's benefits.

The partners suggest that the most important thing when forming a partnership is to be candid and explicit regarding your expectations and ideas. Carefully define exactly what you want the aircraft for. If you want to load the inexpensive VFR airplane you are about to buy with a new stack of IFR avionics as soon as the ink is dry on the bill of sale, by all means, say so before all the partners sign.

4

Make the Numbers Work

Once you understand how to analyze the costs of flying you are ready to find a way to attain your flying objectives on the budget you have available. You should go through as many potential scenarios as it takes to make the numbers work. Be flexible and creative. Varying the number of partners, the number of hours flown per year, the type of aircraft in view of their price and performance, the amount borrowed to buy the airplane, and other variables yield countless permutations, one of which is bound to work for your purse and goals.

This chapter discusses five basic approaches to getting where you want to be. These examples are intended to stimulate your thinking so that you can discover all the alternatives available to you and aggressively and creatively devise your own solutions. The thought process followed is typical of what many pilots go through when considering their partnership options:

- Stage 1–"I want to fly 100 hours a year in an airplane similar to the one I am renting and I have x amount of money per year for flying and y amount of money to buy a share of the plane. How many partners do I need to make it work?"

- Stage 2–"That's interesting, but there are x of us ready to go. What's the most airplane we can afford?

- Stage 3–"Actually, given our flying needs, it would be better to fly an airplane that is a bit higher performance than the one we are renting now. Maybe we should find another partner or two to make this work. How many more partners do we need to afford the higher performance airplane?"

- Stage 4–"I don't really want more than x number of partners because the group would be too unwieldy, but I do want the higher performance aircraft. Maybe I'm overestimating the amount of time I will fly per year. What if I would be willing to fly less? Can I trade quantity of flying for quality of flying on the budget I have?"

- Stage 5–"We've got the number of partners lined up, we know the hours each of us wants to fly, and we know the performance we want from the airplane. Some of the airplanes with the performance we want are too expensive. Is the same performance available for less?"

Let's take a closer look at what the numbers tell us about the options for each of these stages of our thinking. As you go through this process focus on two figures:

- Acquisition cost–the amount of capital required per partner to buy the desired aircraft.

- Annual flying costs–the amount of money it takes per partner to keep and fly the airplane (fixed and operating expenses); look at total hourly expenses and total annual expenses (total hourly expenses x hours flown) per partner to see how these costs compare to your available annual budget.

Bear in mind that if all partners share fixed expenses equally but each pay all of their own hourly operating expenses in full, then, if one partner flies 100 hours per year it will not affect the costs of the partner who flies only 50 hours a year. This is important to note, because it means that partners with different annual flight time goals can become partners without any partner suffering any financial consequences due to the different objectives of the other partners.

Common assumptions made for all the examples to be examined are as follows:

- Hangar/tie-down per month: $85
- State fees per year: $120
- Fuel cost: $2.00 per gallon
- Loan outstanding: no money is borrowed to buy the airplane
- Cost of capital: 7%

A word about the numbers in the examples: particular aircraft types have been in production for many years and are on the used market at a wide range of prices. Acquisition and operating costs fluc-

Fig. 4-1. *The Commander 114 is known for comfort. (Commander Aircraft/Paul Bowers.)*

tuate significantly over the years. The figures used are real-world figures, in effect at the time of writing. However, they might not present current conditions in your geographic area. Perform your own analyses with your own up-to-date numbers.

How Many Partners?

Most pilots begin thinking about a partnership by thinking to themselves that they know which airplane they want, how many hours they want to fly it per year, how much money they have available to buy part of the airplane, and how much money they have to spend on flying every year. How many partners do they need to accomplish their objectives?

Carefully research the information required to do the financial analysis for the airplane you would like and then lay it all out manually or on the computer. Table 4.1 shows the per pilot costs of a 1973 Cessna Skyhawk with a fresh engine and basic IFR avionics for up to six partners.

Note that the capital you need per partner to get this partnership under way drops off significantly with more than three partners. To go it alone you have to cough up $35,000 to buy the airplane and then come up with $7290 every year to fly it. Now look at how dramatically

Cessna 172	AIRCRAFT PARTNERSHIP FINANCIAL ANALYSIS				($/Pilot)	
NUMBER OF PILOTS	1	2	3	4	5	6
CAPITAL INVESTMENT	35000.00	17500.00	11666.67	8750.00	7000.00	5833.33
LOAN AMOUNT	0.00	0.00	0.00	0.00	0.00	0.00
ANNUAL FIXED EXPENSES						
Tiedown/Hangar	1020.00	510.00	340.00	255.00	204.00	170.00
Insurance	800.00	400.00	266.67	200.00	160.00	133.33
State Fees	120.00	60.00	40.00	30.00	24.00	20.00
Annual	550.00	275.00	183.33	137.50	110.00	91.67
Maintenance	1000.00	500.00	333.33	250.00	200.00	166.67
Loan Payments	0.00	0.00	0.00	0.00	0.00	0.00
Cost of Capital (non-cash)	2450.00	1225.00	816.67	612.50	490.00	408.33
Total Fixed Expenses / yr	**5940.00**	**2970.00**	**1980.00**	**1485.00**	**1188.00**	**990.00**
HOURLY OPERATING EXPENSES						
Fuel	16.00	16.00	16.00	16.00	16.00	16.00
Oil	0.50	0.50	0.50	0.50	0.50	0.50
Engine Reserve	6.50	6.50	6.50	6.50	6.50	6.50
General Maint Res	4.00	4.00	4.00	4.00	4.00	4.00
Total Op Exp / hr	**27.00**	**27.00**	**27.00**	**27.00**	**27.00**	**27.00**
TOTAL EXPENSES PER HOUR						
50 Hours flown per year	145.80	86.40	66.60	56.70	50.76	46.80
100 Hours flown per year	86.40	56.70	46.80	41.85	38.88	36.90
150 Hours flown per year	66.60	46.80	40.20	36.90	34.92	33.60
TOTAL ANNUAL EXPENSES						
50 Hours	7290.00	4320.00	3330.00	2835.00	2538.00	2340.00
100 Hours	8640.00	5670.00	4680.00	4185.00	3888.00	3690.00
150 Hours	9990.00	7020.00	6030.00	5535.00	5238.00	5040.00

ASSUMPTIONS:

Aircraft value	35000.00
Insurance/yr:	800.00
Annual:	550.00
Maintenance/yr:	1000.00
Fuel Cost ($/gal):	2.00
Fuel Cons (gal/hr):	8.00
Oil Cons (qt/hr):	0.20
Gen Maint Res/hr:	4.00
Engine Reserve/hr	6.50
Engine MOH Cost:	13000.00
Time Rem. To OH:	2000.00

Table 4.1. *Financial analysis of owning and flying a Cessna 172.*

the costs are reduced if four of you join forces. For each of you to fly 50 hours a year you each need only $8750 to buy the airplane and only $2835 a year to fly it. And to double your flying time to 100 hours a year you need to increase your annual flying budget by only $1350.

Note that as you get beyond three partners the per person savings realized by allowing additional partners to join the group diminishes significantly both in terms of buying the airplane and operating the partnership. Savings on the cost of the airplane is $2917 if the partnership is expanded from four to six people. More significantly, the savings on the annual cost of 50 hours is a mere $495, hardly worth the extra hassles of having to operate the partnership with six people. So if four of you can come up with enough capital to buy the airplane, you will be much better off keeping the partnership small for the marginal increase in total annual operating costs.

The Most Airplane Your Group Can Afford

Another typical approach is for a group of partners to get together without a clear idea of what kind of airplane they want and ask themselves what they can afford. Choose the likely candidates, collect the information on them and lay it all out. You'll quickly see what type of airplane is in line with your budget (but be sure not to lose sight of matching the airplane to your needs).

The airplanes in Table 4.2 are all used 1970s-era aircraft which tend to be good buys if they've been properly maintained. Making these types of comparisons gives you an accurate understanding of what it takes to step up in aircraft categories for a given set of market conditions. In terms of total annual expenses, at the lower end of the

Fig. 4-2. *The Arrow II is a favorite entry-level complex airplane.*

($/Pilot)	Cessna 150	Cessna 172	Piper Archer	Piper Arrow	Beech Bonanza
NUMBER OF PILOTS	3	3	3	3	3
CAPITAL INVESTMENT	6666.67	11666.67	16666.67	20000.00	31666.67
LOAN AMOUNT	0.00	0.00	0.00	0.00	0.00
ANNUAL FIXED EXPENSES					
Tiedown/Hangar	340.00	340.00	340.00	340.00	340.00
Insurance	166.67	266.67	500.00	566.67	733.33
State Fees	40.00	40.00	40.00	40.00	40.00
Annual	133.33	183.33	250.00	283.33	333.33
Maintenance	200.00	333.33	333.33	366.67	833.33
Loan Payments	0.00	0.00	0.00	0.00	0.00
Cost of Capital (non-cash)	466.67	816.67	1166.67	1400.00	2216.67
Total Fixed Expenses / yr	1346.67	1980.00	2630.00	2996.67	4496.67
HOURLY OPERATING EXPENSES					
Fuel	12.00	16.00	16.00	21.00	30.00
Oil	0.40	0.50	0.50	0.50	0.63
Engine Reserve	5.00	6.50	10.00	8.00	10.00
General Maint Res	3.00	4.00	5.00	8.00	15.00
Total Op Exp / hr	20.40	27.00	31.50	37.50	55.63
TOTAL EXPENSES PER HOUR					
50 Hours flown per year	47.33	66.60	84.10	97.43	145.56
100 Hours flown per year	33.87	46.80	57.80	67.47	100.59
150 Hours flown per year	29.38	40.20	49.03	57.48	85.60
TOTAL ANNUAL EXPENSES					
50 Hours	2366.67	3330.00	4205.00	4871.67	7277.92
100 Hours	3386.67	4680.00	5780.00	6746.67	10059.17
150 Hours	4406.67	6030.00	7355.00	8621.67	12840.42

ASSUMPTIONS:	Cessna 150	Cessna 172	Piper Archer	Piper Arrow	Beech Bonanza
Aircraft value	20000.00	35000.00	50000.00	60000.00	95000.00
Insurance/yr:	500.00	800.00	1500.00	1700.00	2200.00
Annual:	400.00	550.00	750.00	850.00	1000.00
Maintenance/yr:	600.00	1000.00	1000.00	1100.00	2500.00
Fuel Cost ($/gal):	2.00	2.00	2.00	2.00	2.00
Fuel Cons (gal/hr):	6.00	8.00	8.00	10.50	15.00
Oil Cons (qt/hr):	0.20	0.20	0.20	0.20	0.25
Gen Maint Res/hr:	3.00	4.00	5.00	8.00	15.00
Engine Reserve/hr	5.00	6.50	10.00	8.00	10.00
Engine MOH Cost:	10000.00	13000.00	15000.00	16000.00	20000.00
Time Rem. To OH:	2000.00	2000.00	1500.00	2000.00	2000.00

Table 4.2. *Analyzing the cost of different aircraft types for a given number of partners.*

aircraft type spectrum it doesn't take much. The annual expenses to fly a Cessna 172 for 50 hours are only $1000 more than it takes to fly a Cessna 150 for the same amount of time. To buy the 172 instead of the 150 is a bit more onerous, requiring an additional $5000 per partner, but even that amount is less than most credit card limits. Considering the amount of extra airplane it buys, the extra capital expense may be well worth it. (Although you would be foolish to put it on a credit card instead of finding a much lower cost source of financing.)

More Airplane for Less with More Partners

The idea that with a group of partners you can afford an airplane that you couldn't afford on your own has an interesting variation: if you are willing to get together with a greater number of partners, you can afford a much more complex airplane for about the same amount of money that you would each spend in a small partnership on a less capable airplane.

In Table 4.3 we compare a two-person partnership in a Skyhawk with a four-person partnership in an Arrow and a six-person partnership in a Bonanza. The per partner cost of purchasing either one of the higher performance airplanes in the larger partnerships is less than the per partner cost of buying the Skyhawk. If you don't mind larger partnerships you need no extra capital to buy a share of a more expensive and capable airplane.

In terms of the annual expenses, the annual fixed expenses per partner are less for the Arrow and the Bonanza. The extra partners adequately compensate for the extra expenses, which are shared equally. The hourly operating costs, which are not shared, are going to be higher per partner for the more capable airplanes. However, they will be somewhat offset by the lower fixed expenses.

In terms of total annual expenses, it costs each of you less to fly 50 hours per partner in the Arrow than it would cost two of you to fly 50 hours each in the Skyhawk. And 50 hours in a Bonanza partnership of six pilots will cost each partner only $586 more than the same time would cost each of two partners in the Skyhawk. That is only $11 per hour more for the Bonanza over the Skyhawk; a bargain you'll not find at any FBO! If you fly 100 hours a year the annual expense difference between the Bonanza and the Skyhawk is $2000 (with increasing flying hours the unshared operating expenses quickly ratchet up the total dollars spent per year), but that is still only $20 more per hour for a Bonanza than a Skyhawk. If you fly 100 hours and want a complex airplane instead of the Skyhawk,

($/Pilot)	Cessna 172	Piper Arrow	Beech Bonanza
NUMBER OF PILOTS	2	4	6
CAPITAL INVESTMENT	17500.00	15000.00	15833.33
LOAN AMOUNT	0.00	0.00	0.00
ANNUAL FIXED EXPENSES			
Tiedown/Hangar	510.00	255.00	170.00
Insurance	400.00	425.00	366.67
State Fees	60.00	30.00	20.00
Annual	275.00	212.50	166.67
Maintenance	500.00	275.00	416.67
Loan Payments	0.00	0.00	0.00
Cost of Capital (non-cash)	1225.00	1050.00	1108.33
Total Fixed Expenses / yr	2970.00	2247.50	2248.33
HOURLY OPERATING EXPENSES			
Fuel	16.00	21.00	30.00
Oil	0.50	0.50	0.63
Engine Reserve	6.50	8.00	10.00
General Maint Res	4.00	8.00	15.00
Total Op Exp / hr	27.00	37.50	55.63
TOTAL EXPENSES PER HOUR			
50 Hours flown per year	86.40	82.45	100.59
100 Hours flown per year	56.70	59.98	78.11
150 Hours flown per year	46.80	52.48	70.61
TOTAL ANNUAL EXPENSES			
50 Hours	4320.00	4122.50	5029.58
100 Hours	5670.00	5997.50	7810.83
150 Hours	7020.00	7872.50	10592.08

ASSUMPTIONS:	Cessna 172	Piper Arrow	Beech Bonanza
Aircraft value	35000.00	60000.00	95000.00
Insurance/yr:	800.00	1700.00	2200.00
Annual:	550.00	850.00	1000.00
Maintenance/yr:	1000.00	1100.00	2500.00
Fuel Cost ($/gal):	2.00	2.00	2.00
Fuel Cons (gal/hr):	8.00	10.50	15.00
Oil Cons (qt/hr):	0.20	0.20	0.25
Gen Maint Res/hr:	4.00	8.00	15.00
Engine Reserve/hr	6.50	8.00	10.00
Engine MOH Cost:	13000.00	16000.00	20000.00
Time Rem. To OH:	2000.00	2000.00	2000.00

Table 4.3. *More partners can buy a more expensive airplane for similar per partner costs than fewer partners have to spend on a less expensive airplane.*

the four-pilot Arrow partnership may be your best bet. It will cost only $318 more per year per partner, or $3.18 per hour.

More Airplane If You Fly Less

As you saw in the previous section, if you are willing to form a larger partnership, you can afford to buy a more expensive and capable airplane for less money than it would take to buy a more modest airplane with fewer partners. You will find, however, that the annual expenses are more if you fly the same amount in the complex machine as you planned to fly in the modest airplane.

Suppose that you and a potential partner have the annual budget to each fly a Skyhawk 150 hours per year, but have a need for a more complex airplane. To find a solution you examine more complex airplanes with more partners in your analysis (Table 4.3), but even with four partners in an Arrow or six in a Bonanza you bust the annual budget of flying the Skyhawk 150 hours with two partners.

There is a solution if you are willing to trade quantity for quality. If you are in a position to fly either complex airplane somewhat less than you were planning to fly the Skyhawk and still meet your flying objectives, you are in business. It doesn't take much.

Table 4.4 compares the cost of 150 hours in a two-person Skyhawk partnership with 100 hours in a four-person Arrow partnership and 50 hours in a six-person Bonanza partnership. The annual costs of flying 100 hours per partner in the Arrow with four partners or 50 hours per partner with six in the Bonanza are both considerably less expensive than for 150 hours in the Skyhawk with two people.

Here's how to find out precisely how much you can fly the Arrow in a larger partnership on the 150-hour, two-person Skyhawk budget: take the total annual expense of the Skyhawk at 150 hours ($7020) and subtract from it the annual fixed expense of the Arrow ($7020 − $2248), which equals $4772. You have $4772 for hourly operating expenses. Divide this amount by the Arrow's operating expense per hour ($37) and you get 129 hours per year. If your flying needs call for an Arrow, you can probably live with 129 hours in it instead of being stuck in a Skyhawk for 150 hours for the same money.

($/Pilot)	Cessna 172	Piper Arrow	Beech Bonanza
NUMBER OF PILOTS	2	4	6
CAPITAL INVESTMENT	17500.00	15000.00	15833.33
LOAN AMOUNT	0.00	0.00	0.00
ANNUAL FIXED EXPENSES			
Tiedown/Hangar	510.00	255.00	170.00
Insurance	400.00	425.00	366.67
State Fees	60.00	30.00	20.00
Annual	275.00	212.50	166.67
Maintenance	500.00	275.00	416.67
Loan Payments	0.00	0.00	0.00
Cost of Capital (non-cash)	1225.00	1050.00	1108.33
Total Fixed Expenses / yr	2970.00	2247.50	2248.33
HOURLY OPERATING EXPENSES			
Fuel	16.00	21.00	30.00
Oil	0.50	0.50	0.63
Engine Reserve	6.50	8.00	10.00
General Maint Res	4.00	8.00	15.00
Total Op Exp / hr	27.00	37.50	55.63
TOTAL EXPENSES PER HOUR	46.80	59.98	100.59
	150 hr/yr	100 hr/yr	50 hr/yr
TOTAL ANNUAL EXPENSES	7020.00	5997.50	5029.58

ASSUMPTIONS:	Cessna 172	Piper Arrow	Beech Bonanza
Aircraft value	35000.00	60000.00	95000.00
Insurance/yr:	800.00	1700.00	2200.00
Annual:	550.00	850.00	1000.00
Maintenance/yr:	1000.00	1100.00	2500.00
Fuel Cost ($/gal):	2.00	2.00	2.00
Fuel Cons (gal/hr):	8.00	10.50	15.00
Oil Cons (qt/hr):	0.20	0.20	0.25
Gen Maint Res/hr:	4.00	8.00	15.00
Engine Reserve/hr	6.50	8.00	10.00
Engine MOH Cost:	13000.00	16000.00	20000.00
Time Rem. To OH:	2000.00	2000.00	2000.00

Table 4.4. *Access to a higher performance airplane can be achieved for less per pilot compared to a lower performance airplane if the pilots are willing to trade quantity for quality by flying less.*

If you do a similar analysis of the six-person Bonanza partnership compared to the 150-hour, two-person Skyhawk budget, you'll find that you can fly the Bonanza 88 hours a year for the same money [($7020 - $2132)/$55.50 per hour = 88 hours]. This too may be the perfect trade-off if your goals require the Bonanza experience and you can deal with the larger partnership.

The Performance You Want, for Less

You may be partial to a particular airplane and a particular group of partners, but if you are on a budget, your group may not be able to afford the airplane you've set your sights on. Do not despair, for it may be easier than you think to meet the flying objectives you established for yourselves.

Don't think of the airplane as a particular type. Think of it as performance. When you buy an airplane pragmatically you are buying it for what it can do, how it can perform. The performance you seek is dictated by your flying objectives. So forget airplane type. Make a list of the performance criteria that meet your objectives and scour the performance statistics to find the least expensive used airplane that meets your requirements (see Table 4.5).

A brand new 1997 Cessna Skyhawk that rolls out the factory door for $140,000 will take you and three friends just about as fast and as far as a well-maintained 1969 Cessna 172 with a fresh engine that can be had for $30,000, because basically the two airplanes are almost identical. The new Skyhawk's interior is pretty, but not $110,000 pretty! It is much better soundproofed, but you can "upgrade" the 1969 model with two noise canceling headsets for only $1000.

If you prefer something more recent than 1968 and like low-wing airplanes, how about a 1980s Piper Warrior? It can be had for a little

Fig. 4-3. *The Maule is a true bush plane.*

more than the 1969 Skyhawk and its operating costs are basically the same.

Note that in buying performance, the more sensitive variable is the purchase price of the airplane. The operating costs tend to be similar if performance is similar. Generally maintenance is more expensive for older airplanes, but if you buy carefully this is not necessarily true. If you make sure that expensive recurrent items have been done recently, that the engine was well overhauled and has a lot of time remaining on it, and you do your research very carefully to avoid airplane types that have proven to be maintenance dogs over the years, you should have no surprises.

Keep the performance approach in mind when you and your partners are considering the most airplane you can afford. A 1970s Grumman Tiger, for example, is one of the best kept secrets in the

($/Pilot)	1960's Cessna 172	1980's Piper Warrior	1997 Cessna 172
NUMBER OF PILOTS	3	3	3
CAPITAL INVESTMENT	10000.00	15000.00	43333.33
LOAN AMOUNT	0.00	0.00	0.00
ANNUAL FIXED EXPENSES			
Tiedown/Hangar	340.00	340.00	340.00
Insurance	250.00	316.67	1000.00
State Fees	40.00	40.00	40.00
Annual	150.00	150.00	150.00
Maintenance	366.67	333.33	200.00
Loan Payments	0.00	0.00	0.00
Cost of Capital (non-cash)	700.00	1050.00	3033.33
Total Fixed Expenses / yr	1846.67	2230.00	4763.33
HOURLY OPERATING EXPENSES			
Fuel	16.00	16.00	16.00
Oil	0.50	0.50	0.50
Engine Reserve	6.50	6.50	6.50
General Maint Res	6.00	5.00	3.00
Total Op Exp / hr	29.00	28.00	26.00
TOTAL EXPENSES PER HOUR			
50 Hours flown per year	65.93	72.60	121.27
100 Hours flown per year	47.47	50.30	73.63
150 Hours flown per year	41.31	42.87	57.76

Table 4.5. *The same performance can be bought for less.*

TOTAL ANNUAL EXPENSES

50 Hours, Own	3296.67	3630.00	6063.33
100 Hours, Own	4746.67	5030.00	7363.33
150 Hours, Own	6196.67	6430.00	8663.33

ASSUMPTIONS:	1960's Cessna 172	180's Piper Warrior	1997 Cessna 172
Aircraft value	30000.00	45000.00	130000.00
Insurance/yr:	750.00	950.00	3000.00
Annual:	450.00	450.00	450.00
Maintenance/yr:	1100.00	1000.00	600.00
Fuel Cost ($/gal):	2.00	2.00	2.00
Fuel Cons (gal/hr):	8.00	8.00	8.00
Oil Cons (qt/hr):	0.20	0.20	0.20
Gen Maint Res/hr:	6.00	5.00	3.00
Engine Reserve/hr	6.50	6.50	6.50
Engine MOH Cost:	13000.00	13000.00	13000.00
Time Rem. To OH:	2000.00	2000.00	2000.00

Table 4.5. *(Continued.)*

used aircraft market. It will give you Piper Arrow performance on a 180-hp engine without the complications and maintenance costs of the Arrow's retractable gear and constant-speed propeller. It is more economical than a comparable Arrow and sells for 20% to 30% less.

When you consider aircraft models that have been in production for many years, research them very carefully. Not all model years by the same name are similar airplanes, and some model years of some aircraft have potential problems that may make them poor candidates for meeting your flying objectives. Everyone knows about the fiasco of equipping the Skyhawk with the camshaft and valve train eating Lycoming O-320-H engine between 1977 and 1981. But you should also be aware, among other things, that until 1968 the Skyhawk had a 145-hp Continental engine.

The examples in this chapter demonstrate the infinite permutations you can consider to get the best solution for your flying needs, but they have barely scratched the surface. Keep an open mind, consider every angle, run projection after projection, be creative in making comparisons, and make the numbers work for you.

5

The Co-ownership Agreement

The co-ownership agreement is the aircraft partnership's foundation. Note that I don't call it a partnership agreement because, for liability reasons, the preferable technical legal form of an airplane partnership is a co-ownership. A joint ownership that is a partnership in the legal sense implies that each partner is responsible for the actions of the others. In a legal co-ownership, however, the implication is that each co-owner is responsible only for his or her own actions.

The best liability protection is provided by incorporation, which is discussed below. If you incorporate, the agreement governing your group would be similar to the co-ownership agreement, but would be called the corporation's by-laws. Regardless of what option you choose, to keep things simple I will refer in this chapter to the agreement that governs it as the co-ownership agreement and the members of your group as co-owners. Please note that this book does not give specific legal advice. You should consult your lawyer to help you structure your group to meet your specific needs.

The co-ownership agreement spells out the rights and obligations of each co-owner. It sets the rules by which the co-ownership is established, operated, and dissolved. A good co-ownership agreement clearly defines how to handle the day-to-day issues of a co-ownership, such as scheduling, insurance, expenses, and decisions. It also guides the co-owners through dealing with less common events, such as disputes or the departure of a co-owner.

A surprisingly high number of pilots about to enter into aircraft co-ownership see no need for a co-ownership agreement. Are they not joining forces, they ask, in a common cause with like-minded fliers, often best friends, among whom never a harsh word has been spoken? What could possibly go wrong? Plenty!

The mistake these pilots make is that they fail to consider the unexpected, the unpredictable. Co-ownerships with no written rules can quickly experience internal friction once the initial euphoria of owning an airplane wears off. An idyllic first flight with another co-owner on that silky summer morning is fast forgotten when you are freezing your Jeppsens off in the dead of winter, staring at that same co-owner's parked car on the ramp where you expected to find the airplane.

The co-ownership can also be threatened from the outside. Your co-owner may, indeed, be the nicest, most understanding and generous person; so nice, in fact, that on a whim he lent the airplane to his Uncle Harry, who decided to land on the grass at the local fly-in, stuck a wheel in the mud, and took out a Stearman, a Glasair, and a brand new Bonanza as he veered off the runway. Who do you think the mob will be after? Not just Uncle Harry. You need a good co-ownership agreement.

Should You Get a Lawyer?

A lawyer's advice in putting together a co-ownership agreement is highly recommended. Only a lawyer can advise you effectively on what legal form of existence your group should take to best protect your interests given the applicable laws in your state. There are also well-established legal conventions for formulating the rules that make co-ownerships work. Lawyers get paid to bring to your attention all the items that should be covered in a co-ownership agreement and advise you on how to handle them. They can suggest easy solutions for scenarios you never even envisioned. In a worst-case dispute their carefully crafted co-ownership agreements may well shed enough light on the problem to resolve it amicably, without having to resort to the legal system.

The trouble with lawyers is that they are expensive and many of them have a tendency to err on the side of caution, devising voluminous legal tomes that go way beyond your simple requirements. This, however, should not deter you from using the services of a lawyer for your co-ownership agreement. Just do your homework and develop a good idea of what you want. Spell everything out to your lawyer clearly, and accept or reject suggestions decisively. One reason why lawyers so often put on a dog and pony show is because they are searching in good faith for what an ill-informed, wishy-washy client really wants.

The best way to keep legal costs down yet benefit from a lawyer's advice is to draft a co-ownership agreement with your co-owners in layman's language. Have the lawyer review your draft and put it in final form for your approval. If you and your co-owners are well prepared, the lawyer's time spent on your co-ownership agreement should rarely exceed an hour or two. And remember, the agreement you get for your share of a rather modest fee protects an investment worth tens of thousands of dollars.

Should You Incorporate?

Incorporation is the most effective way to limit your liability for the actions of your co-owners and, under certain circumstances, your own actions. Incorporation is always your safest option, but if you have any significant assets to protect, such as a house in which you have a lot of equity or a large investment portfolio, it is your only option.

When your group is incorporated (registered with the appropriate authorities as a corporation), it becomes a separate and distinct legal entity in its own right. The corporation may enter into commitments on its own behalf and has the right to own assets. In the case of a co-operative flying venture, the corporation's sole asset is the airplane. You and the other members of the group own a share in the corporation rather than shares of the airplane directly.

Liability protection

As a legally incorporated entity, the corporation, rather than its shareholders, is legally liable for any commitments made by it, and for any consequences of the activities performed under its auspices. If a member of an incorporated flying group causes damage or injury while operating the group's aircraft, the victim's legal recourse is limited to the corporation and, under certain circumstances, such as personal negligence, the shareholder causing the problem. The rest of the shareholders are shielded from liability. In case of a judgment against the corporation, it is liable to the extent of its insurance coverage and may also lose its assets (the airplane) to satisfy the claimant, but the shareholders' personal assets are off limits.

The limited liability company (LLC)

Traditionally incorporation was unattractive to small groups because of the onerous administrative burdens, including the voluminous corporate tax filings required to maintain it. However, in recent years incorporation has become a very easy procedure, largely to make it possible for small businesses to function in the simple manner of sole proprietorships or co-ownerships, yet benefit from the liability protection of the corporation.

The first such simplified form of incorporation was the so-called subchapter-S corporation. As long as a handful of conditions, such as a limit on the number of shareholders, is met, a company may elect to be an S corporation. Its shareholders add a simple subchapter-S tax form to their personal income tax forms (and may elect to treat corporate income as personal income). If the corporation made no money during the year (as would be the case with a cooperative flying venture), a modest annual minimum tax is payable to maintain its corporate existence.

Recently a new form of incorporation has come into being in all 50 states, the *limited liability company*. The conditions to be met to form an LLC are less restrictive than those of the S corporation, but it retains all the benefits of the S corporation. Its shareholders can function as a co-ownership yet retain the liability protection of a corporation.

Maintaining proof of corporate existence

If you do incorporate, it is important to perform certain basic functions to prove that your entity is a bona fide corporation. You must hold annual meetings, keep a minimal amount of corporate records, and make the required tax and corporate filings. Failure to do so could invalidate the legal existence of your corporation and result in the loss of liability protection.

The Typical Co-ownership Agreement

All co-ownership agreements should cover certain essential points. Where you have some flexibility is in the degree of detail in which you decide to address some of the issues. You can be quite vague, or very specific, depending on the issue, your co-ownership's circumstances, and legal considerations. Here is where a lawyer can be very helpful. The best approach is to cover each point in sufficient detail to leave no ambiguity about information and to provide a clear mechanism for resolving any conflicts that may arise.

The basic issues that the average co-ownership agreement should address include

1. the asset, the co-owners, and ownership
2. use of the aircraft
3. legality clause
4. persons authorized to fly
5. decision making
6. expenses
   fixed expenses
   operating expenses, including engine and maintenance reserves
   payments
   delinquencies
   statement of accounts
7. storage
8. maintenance
9. insurance
10. loss or damage of the aircraft
11. death or disability of a co-owner
12. departure of a co-owner or termination of the co-ownership
13. annual statement of value
14. collateralization exclusion
15. communications
16. arbitration
17. amendments

These issues are discussed in detail below. Bear in mind that not all points may be useful to all co-ownerships and the degree of detail in your agreement will depend on your preferences and circumstances. Use this information as a set of guidelines, a point of departure for formulating your own co-ownership agreement, tailored to suit your specific circumstances. Let's go through the sample agreement point by point (appendix 2 contains the agreement without the explanatory text).

This document memorializes a co-ownership agreement between:

> *John Smith*
> *of 22 Main Street, Anytown, MA*
> *and*
> *Joe Doe*
> *of 93 Fox Road, Smallville, MA*

(referred to hereafter as the "co-owners").

This is the agreement's title page. It identifies the legal name and addresses of the co-owners.

> *1. The Asset, the co-owners, and ownership.*
> *The co-owners own a 1978 Piper Cherokee Archer aircraft,*
> *model PA 28-181, serial number 28-6947032, registered as*
> *N4968X (referred to as the "aircraft"). Each co-owner owns*
> *an undivided fifty percent (50%) interest in the aircraft.*

This is the opening statement of the co-ownership. It identifies the co-ownership's asset (the airplane) in as much detail as possible. Make and model according to the manufacturer's official designation should be identified, the manufacturer's serial number should be included, and the FAA registration number (N number) should be listed. If you don't have a specific aircraft when you are putting together the co-ownership agreement, fill in the aircraft data later.

This section also identifies the ownership share of each co-owner (customarily by percentage). It is important to state that ownership is "undivided," expressing each co-owner's share as a percentage of the total aircraft. This means that all co-owners own a percentage of the entire airplane; the airplane is not divided into identifiable pieces among the co-owners. One co-owner does not own the propeller, another the fuselage, and another the engine and the instruments. This may seem like legal nit-picking, but it eliminates exploiting legal technicalities should a co-ownership dispute get out of hand ("I'm taking the wings home, they are mine!").

If you choose to operate the co-ownership under a group name, this section is a good place to identify that name. Alternatively you can have a separate clause identifying the name.

A common question asked about co-ownerships is what happens to the co-ownership agreement and the title document filed with the FAA when one co-owner sells his or her share to someone outside the co-ownership. One option is to list all co-owners individually in both the agreement and the FAA title document. In this case the agreement must be amended when a co-owner's share is sold, and the FAA title must be refiled listing the new co-owner.

Another choice is to hold title under a co-ownership name via transferable shares so that the FAA title does not have to be refiled when a share is sold. The shareholders may be specifically listed by name in the agreement, which would require amendment when a share is sold to replace the seller's name with the buyer's name.

Fig. 5-1. *A used Mooney 252 can be had for a fraction of the price of its new equivalent.*

Alternatively separate paper shares may be held by each shareholder, which can be transferred by signature to a new shareholder when the share is sold. This mechanism of share transfer is standard if you go the incorporation route. The corporation owns the airplane, and a departing shareholder signs his or her shares over to the buyer of the shares without any effect on the title document filed with the FAA.

If you do not incorporate and do not anticipate frequent turnover in co-owners, perhaps the simplest solution is to name each co-owner in both the agreement and the FAA title document, and make amendments and title refilings as and when appropriate.

You should be particularly careful in any case to devise some form of consent by all co-owners to the sale of any co-owner's share (for more details, see "12. Departure of a co-owner or the termination of the co-ownership" later in this chapter). This is another area where a lawyer can be most helpful in devising a solution to best suit your particular needs and ensure that all the documentation is in the proper legal form.

  2. Use of the aircraft.
    Each co-owner shall have full use and benefit of the aircraft on a schedule to be mutually agreed upon. No co-owner will use (or permit the use of) the aircraft under conditions which could have the effect of voiding the hull or liability insurance on the aircraft.

This section lays out who can use the airplane, according to what schedule, and under what conditions. It can also explicitly forbid illegal uses. The section can be as specific or as general as the co-owners prefer. All issues may be handled under one clause, or you may have separate clauses such as "Scheduling" and "Persons authorized to fly."

  Scheduling can be addressed in varying degrees of detail depending on the co-ownership's needs and circumstances. For small, flexible co-ownerships the wording "a schedule mutually agreed upon" will usually suffice. For larger co-ownerships with periods of peak demand for the airplane (mostly on weekends), more explicit scheduling rules may be required.

  3. Legality clause.
    All co-owners agree to operate the aircraft in accordance with Federal Aviation Regulations and state and federal laws.

This clause requires all persons authorized to fly the airplane to agree to fly it only in compliance with all legal requirements. It is a good catchall clause obligating the co-owners to have current medicals, get the appropriate weather briefings, and meet other requirements for operating the airplane without having to separately state these requirements in the agreement. This section may be a subclause of "2. Use of aircraft," but should be explicitly stated.

  4. Persons authorized to fly.
    No person is authorized to fly the co-ownership's aircraft other than the co-owners and appropriately authorized flight instructors providing instruction to co-owners.

In general you may want only the co-owners to fly the airplane, with the exception of certified flight instructors giving instruction to co-owners. However, there may be other occasions when a co-owner may benefit from a non-co-owner acting as pilot in command. Examples are an A&P mechanic who barters his services for flight time, a co-owner whose spouse is a pilot but for whatever reason is not a member of the co-ownership, or a co-owner who does aerial photography from time to time and needs a qualified photo pilot.

Be sure to explicitly state the qualifications and experience required of non-co-owners allowed to fly the co-ownership's airplane. Be sure to include licenses held and ratings such as "authorized flight instructor" and "commercial pilot with instrument rating" as well as general flight experience, experience in type, and check ride requirements. To minimize the chance of misadventure you may want to set considerably higher standards for non-co-owners compared to co-owners.

It is extremely important to check your insurance policy (as well as every policy renewal) for what it says about non-co-owners being allowed to fly the co-ownership's airplane. There may be severe insurance restrictions on such flying. Non-co-owners may be completely forbidden from flying the airplane. They may have to be specifically named on your policy, and there may be additional qualifications and experience requirements. To permit non-co-owners to fly your airplane on your insurance policy, you may have to incur significant additional insurance expense.

5. Decisions.

All decisions except decisions related to the day-to-day operation of the aircraft as pilot in command will be made by consensus.

Decisions, decisions. How will the co-ownership make them to keep all the co-owners happy all the time? The key question is whether unanimous agreement should be required or if some form of majority should suffice. Lawyers will often favor a two-thirds majority vote to make a decision binding, but they tend to forget that a co-ownership for private flying is not business or politics, but fun, and should strive to minimize the chance for hard feelings. Therefore, unanimous agreement to make a decision binding is highly recommended.

6. Expenses.

6.1. Fixed expenses.

All fixed expenses arising from the ownership of the aircraft, including but not limited to tie-down or hangar charges, insurance charges, property taxes, and the annual inspection shall be borne equally by the co-owners.

6.2. Operating expenses.

All operating expenses arising from the operation of the aircraft, including but not limited to fuel and other expendables, and maintenance and overhaul reserves shall be borne in their entirety by the

Fig. 5-2. *A Malibu Mirage may make a lot of financial sense for a business partnership.*

partner operating the aircraft. The co-owners agree to establish mutually acceptable hourly rates for engine overhaul and maintenance reserves at a level that will ensure the accumulation of projected maintenance and engine overhaul costs. The hourly rate for these reserves may be periodically readjusted by consensus as required.

6.3. Payment of expenses except fuel.

The co-owners agree to open a joint co-ownership checking/savings account for making payments related to the aircraft. The co-owners further agree to designate one of their number for a mutually acceptable term as Treasurer, to operate the checking/savings account and maintain a careful record of payment of the aircraft expenses.

Each co-owner agrees to forward payment for his or her share of all fixed expenses and operating expenses, except fuel, in a timely fashion to the Treasurer, who will deposit these payments into the co-ownership account and make payments on behalf of the co-ownership related to the aircraft out of the account.

Except as stated otherwise in this agreement, no expense in excess of $200 may be made on behalf of the co-ownership without prior approval of all co-owners.

The co-owners agree to meet, review, and reconcile all aircraft expenditures at least every three (3) months, when the

Treasurer will provide statements for the period covered to all co-owners.

6.4. Assessments.

The co-owners agree to make assessments as necessary to cover unexpected maintenance and engine overhaul expenses if the funds in the account to cover such expenses fall short. Each co-owner will pay a percentage share of the assessment equal to the percentage share of that co-owner's flight time of the total flight time completed in the aircraft by the co-owner-ship since such maintenance or engine overhaul was last done on the aircraft.

6.5. Payment of fuel expenses.

Each co-owner will pay his or her fuel expenses directly to a fuel supplier of his or her choice. Each co-owner will leave the aircraft refueled to the tabs after every flight.

6.6. Delinquencies.

A co-owner whose payments are 60 days past due loses all flight privileges until the delinquency is cured. A co-owner whose pay-ments are 90 days past due and equal or exceed $2000 may be required at the option of the other co-owners to sell his or her share of the co-ownership to the other co-owner(s) or a third party at a valuation of the annual stated value.

Expenses should be addressed in great detail. The handling of ex-penses can be a big source of friction for a co-ownership. A thought-ful agreement will go a long way toward forestalling later headaches. The two main points you need to address are how divide expenses among the co-owners in principle, and how the physical payment of expenses should be handled and recorded.

The first step is to state what items the co-ownership considers to be expenses by simply listing them. Also include the "including but not limited to" escape language, in case you face an expense at a later date that you did not list.

Next you need to spell out each co-owner's share of expenses. In deciding how to determine each partner's share of the expenses, it is helpful to think in terms of fixed and operating expenses. To recap, fixed expenses are expenses you have to pay regardless of the amount of time the airplane is flown, such as tie-down fees, the an-nual inspection fee (excluding maintenance requirements uncovered by the inspection), and any taxes and fees; operating expenses are the expenses directly associated with flying the airplane, such as fuel, oil, and the amount you decide to charge per hour to rebuild an en-gine and for the maintenance reserve.

There are two schools of thought on sharing expenses. One holds that fixed expenses should be shared equally by all co-owners, regardless of the amount of flying each co-owner does, because these expenses are the cost of owning and preserving the investment; operating expenses should be paid individually by each co-owner per flying hour, because this represents a benefit only to the co-owner using the airplane and pays for the wear and tear put on the airplane by that co-owner.

The other school of thought holds that if there is a substantial difference in the amount of hours flown by the co-owners, fixed expenses should also be divided according to the hours flown because the co-owners flying more get more "use" out of the fixed expenses and should pay for it. I disagree with this logic, because these expenses do not vary with the number of hours flown. The amount to be paid by the co-ownership is the same regardless of how little or how much any co-owner flies, and should therefore be shared equally by all co-owners. Any "use" is taken care of by the hourly engine and maintenance reserve charge, paid separately by each co-owner per hour flown. Review both arguments carefully and make up your own mind. What is important is that your co-owners reach the same conclusion. A fixed expense where an argument may be made for proportional sharing is insurance in cases where there is a big difference in the flight experience levels among the co-owners. Insurance premiums vary greatly with the level of flight experience. A private VFR pilot with 250 hours will usually have to pay a much higher premium than a commercial pilot with an instrument rating and 1500 hours. If the two are co-owners in an airplane, their insurance premium will be set at the rate applicable to the pilot with the lower level of experience. If they split the premium equally, the commercial pilot will subsidize the insurance expense of the private pilot. The way to determine the fair share for each pilot is to get two quotes, one based on the real experience levels of both pilots, and one assuming that both pilots have the experience level of the more experienced pilot. The proportional share for each pilot is then adjusted in favor of the more experienced pilot by the difference in the two quotes.

Unforeseen maintenance and engine overhaul expenses are always a possibility. If the engine needs major work or has to be overhauled before its time there may not be enough money in the account to cover the expense. In this case, applying the concept of "share of wear and tear," each co-owner should be assessed for the extra amount required in proportion to that co-owner's share of the total amount of flying time in the aircraft by the co-ownership since the maintenance item or engine overhaul was last completed on the aircraft.

Having defined how to allocate expenses within the co-ownership, you have to devise a system for making payments and keeping accounts. Depending on the number of co-owners and the relationship between them, these arrangements can range from informal occasional settlements between the co-owners to a system of accounts and comprehensive payment rules controlled by one of the co-owners acting as treasurer. Whatever your choice, the issues you need to deal with are how payments will be made for bills presented to the co-ownership, how hourly operating expenses such as fuel will be handled, how engine and maintenance reserves will be collected, how late payments by a co-owner will be handled, how unforeseen big expenses will be paid, and how often and in what form statements will be provided to the co-owners.

Even the simplest arrangement should spell out certain requirements designed to handle any unforeseen problems that may arise. At a minimum, the agreement should be worded to set time limits for payment of a co-owner's share of a bill. It should set a maximum dollar limit on co-ownership expenses which a co-owner may incur on behalf of the co-ownership without consulting the other co-owners. The agreement should also require co-owners to meet periodically (quarterly or at least semiannually) to reconcile accounts.

Larger, more formal co-ownerships or co-ownerships with especially exacting personalities should build on this simple structure to create a system for making payments and keeping accounts tailored to specific needs and circumstances. Two bank accounts may be established for the co-ownership, one to build the engine reserve, and one to serve as the operating account. A per flying hour general maintenance reserve independent of the engine reserve can be paid into the operating account.

A co-owner should be appointed as treasurer. He or she should be sent all the payments by the other co-owners for deposit into the co-ownership accounts as appropriate (the accounts should be managed by the treasurer, but all co-owners should have signing authority, and expenditures or withdrawals above a certain amount could require two signatures). In turn, the treasurer should be required to send quarterly statements to all co-owners (copies of the bank statements for the period and the co-ownership's income/expense log clearly identifying each item should suffice).

The co-owners should have the right to inspect the books at any time, and a CPA may be retained once a year for a modest fee to independently review the records. The annual CPA review is an especially recommended sanity check for co-ownerships in which no co-owner has much financial experience.

Fuel may be most simply handled by requiring that the tanks be topped off after each flight. Some co-ownerships devise a system of fuel credits and debits based on hours flown, on the theory that it is not efficient to refuel every time you fly around the patch for 30 minutes. Tracking such fuel debits and credits may prove to be more trouble than it is worth, so consider it carefully before committing to it.

For expenditures beyond a certain amount, approval of the entire co-ownership should be required.

It may be worth recognizing in the agreement a co-owner's right to install equipment entirely at his or her own expense upon approval by the other co-owners. There are instances when a well-to-do co-owner is willing to spring for, say, an expensive global positioning system (GPS) when the others simply don't have the money.

Payment delinquencies can be particularly annoying in a co-ownership because they can threaten the smooth operation of the aircraft. Time limits for prompt payment should be established. The consequences for late payment, such as loss of flying privileges, should be spelled out and strictly enforced. A mechanism should be devised to force the delinquent co-owner out of the group if the delinquency exceeds a certain limit set by the co-ownership.

7. Storage.

The aircraft shall be hangared at an airport mutually acceptable to the co-owners.

Away from home the aircraft may be tied down. It is the responsibility of each co-owner using the aircraft to insure that the aircraft is locked and appropriately secured on the tie-down. When tied down overnight, the aircraft's canopy cover will be put in place.

The terms of storage may be worth spelling out, especially in a large co-ownership. Among finicky co-owners it is also best to clarify storage requirements away from home. In smaller co-ownerships these issues may be taken care of by verbal agreement until a problem arises that requires a written standard.

8. Maintenance and repairs.

The co-owners agree to have the aircraft maintained in accordance with the manufacturer's recommendations and all applicable laws. All maintenance will be appropriately documented. Except for emergency field repairs, all maintenance will be conducted by a maintenance facility chosen by the co-owners' mutual consent.

Major scheduled maintenance and engine overhaul will be scheduled sufficiently in advance to preempt inconveniencing co-owners.

All maintenance in excess of $500 must be approved by the co-owners' mutual consent.

Co-owners may perform owner maintenance as permitted by FAR 43.3(g), but must have all such work inspected by a mutually acceptable A&P mechanic.

Emergency field repairs away from home may be performed only by an FAA authorized maintenance facility. Such repairs in excess of $500 should first be discussed with the other co-owners,

Fig. 5-3. *The co-ownership agreement should address adding equipment.*

except when a co-owner cannot be reached in spite of all reasonable efforts.

This clause obligates the co-owners to properly maintain the airplane, but it is worded loosely enough to provide them sufficient flexibility to mutually agree upon the type of maintenance arrangements they desire. This could range from a freelancing A&P mechanic through a repair facility that allows owners to participate in maintenance, to the fanciest and most expensive maintenance companies.

Limits should be placed on the cost of maintenance that may be performed without the approval of all co-owners to prevent financial surprises.

Also this is the place to specify if any owner maintenance will be allowed (recommended only if everyone is very comfortable with the idea).

9. Insurance.

The co-owners agree to maintain full hull and liability insurance at a mutually acceptable level to be determined at the beginning of each insurance period and subject to review at any time at the request of a co-owner. The value of the aircraft for insurance purposes will be the value according to the current Aircraft Blue Book Price Digest.

This clause specifies how the co-owners are to determine what type of insurance coverage to get and the amounts of hull and liability coverage to carry on the airplane. The clause can be as broad or as specific as the co-owners would like. "A level of coverage mutually agreed upon" will keep things simple. Alternatively, for hull insurance coverage, an option is to specify current market value as defined above. For liability insurance, specific levels of coverage may be listed based on the co-owners' levels of net worth threatened by potential liability suits.

If you allow non-owners to fly the airplane, specify the coverage required.

10. Loss of or damage to the aircraft.

In the event of damage to the aircraft for any reason, the co-owner using the aircraft at the time the damage is incurred will be responsible for payment of the uninsured portion or deductible of the claim. If the aircraft is damaged under circumstances where it was not in use by one of the co-owners, then the uninsured or deductible portion of the claim shall be borne equally.

This is an important technical section spelling out the financial responsibilities of the co-owners in case of loss of or damage to the aircraft. The language covers three scenarios: loss or damage when not in use, and as a result of legal use are explicitly addressed, loss or damage, and as a result of illegal use is implied.

When illegality is not an issue, the question of financial responsibility for the loss of or damage to an insured aircraft is reduced in practical terms to who will pay the deductible. When a properly parked and tied down aircraft is lost or damaged when not in motion, no one co-owner is responsible, so all should bear equal responsibility and pay an equal share of the deductible. When an aircraft is damaged while in motion (in use by a co-owner), the co-owner in command should be liable for the deductible. In case of illegal use which negates insurance coverage, the co-owner in command should be financially liable for the entire value of the aircraft. Should the co-owner be lost with the aircraft during illegal use which negates insurance coverage, the surviving co-owners may collect from the estate, if there is anything to collect.

The rules for financial responsibility by nonowners for loss of or damage to the co-ownership's aircraft while in use by a nonowner, should the co-ownership allow such use, are similar to the rules applied to co-owners. However, nonowners may have very few assets from which to make restitution in case of loss or damage as a result of illegal use.

Nonowners who have renter's insurance may be out of luck, because unless a co-ownership is commercially insured it is not allowed to rent its airplane.

11. Death or disability of a co-owner.

In the event of the death or disability of a co-owner, the remaining co-owner may purchase the other co-owner's interest for fifty percent (50%) of the most recent agreed-upon fair market value pursuant to paragraph 13 and exhibit A.

In the sad case of a co-owner's death or disability it is important for the co-ownership to protect itself from disposal of the deceased or disabled co-owner's shares by the estate or guardian on terms unfavorable to the co-ownership. Estates or guardians may wish to liquidate the co-owner's share at the highest possible value in the shortest amount of time. Thus, if the co-ownership is prepared with a reasonable and speedy option, all may work out for the best for everyone.

The first order of business is to establish the co-ownership's right of first refusal at a fair price. This price may be established according to the method used in the annual statement of value, an important reason for choosing a credible method. Should this be acceptable to the estate or guardian, all the co-ownership has to do is come up with the money or find an acceptable replacement co-owner. The co-ownership should have a reasonable time (say, 60 days) to come up with a replacement co-owner.

A second line of defense is to specify that whoever the estate sells to must be acceptable to the co-ownership, and that the value of the buyer's share will not be worth more than the current market value (as defined in the annual statement of value) of the share of the co-owner being replaced.

If no solution can be found, the last resort is to dissolve the co-ownership.

> *12. Departure of a co-owner or the termination of the co-ownership.*
>
> *This co-ownership may be terminated by written notice given to the other co-owner at the address indicated above. Such notice shall also state whether the co-owner electing to terminate desires to buy the aircraft outright or sell his interest to the remaining co-owner. The co-owner receiving such notice shall then inform the other co-owner as to his desire to buy or sell the aircraft. In this situation, the following rules will apply:*
>
> *12.1. If both co-owners desire to sell the aircraft, it shall be sold to a third party and the proceeds divided equally (following a final accounting between the co-owners).*
>
> *112.2. If one co-owner desires to sell and one co-owner desires to buy, the purchase price shall be fifty percent (50%) of the most recent agreed upon fair market value pursuant to paragraph 13 and exhibit A.*
>
> *112.3. If one co-owner desires to sell and the other co-owner desires to maintain ownership through a new co-ownership with a third party, the co-owner desiring to sell will give the other co-owner 60 days to find a new co-owner.*
>
> *112.4. If both co-owners desire to buy the aircraft, then both co-owners shall submit simultaneously to an independent umpire a sealed bid representing the highest figure that he will pay to purchase the other co-owner's interest in the aircraft. The independent umpire will review both figures, determine which co-owner has offered the higher figure, and the aircraft shall be sold to the co-owner making the higher offer at the amount of that offer.*

A co-owner wishing to leave the co-ownership should be required to give written notice. The departing co-owner should expect fair value for his or her share, and in turn should cooperate to the extent possible with the remaining co-owners to preserve the co-ownership on the most desirable terms.

The remaining co-owners should have the right of first refusal for a reasonable period (say 30 to 60 days) at the current market value. They should also be given the option of finding a replacement co-owner during this period.

The seller should have the right to propose replacement co-owners for the unanimous approval of the remaining co-owners. If a buyer is found, the remaining co-owners should have the option of matching the offer in lieu of accepting the buyer.

The share of the joining co-owner should equal in percentage the departing co-owner's share of the co-ownership regardless of the dollar amount for which it was sold. The seller may have decided to sell at a discount to current market value, or may have been lucky enough to sell at a premium, but for the co-ownership to remain intact on equal terms, the only figure that counts is percentage share, and this should be acceptable to the buyer.

If the co-owners cannot reach agreement, the co-ownership should be dissolved and the aircraft sold to the highest bidder following exposure to the market for a reasonable length of time.

If more than one co-owner wishes to buy the entire airplane, the highest bidder wins on a sealed bid. In such an eventuality the co-ownership should agree on a minimum acceptable bid.

> *13. Annual statement of value.*
>
> *The co-owners agree that as of the date of this agreement, the aircraft's fair market value is $50,000. The co-owners agree that they shall prepare an addendum to this agreement signed by each of them stating the fair market value of the aircraft (to be defined according to a mutually acceptable method) every six months commencing from the date of this agreement in form attached as exhibit A.*

Airplane values fluctuate over time, and the value of a particular airplane may be increased by improvements, such as new avionics or an overhauled engine. It is important for the co-ownership to periodically establish the current market value of the airplane for a variety of reasons. Establishing the current value will determine the amount of insurance to carry on the airplane. Should a co-owner decide to leave, the current value will give the remaining co-owners an idea of what to reasonably offer for the departing co-owner's share.

This clause should define how fair market value is established and how often it is determined. A common way to assess market value is to use the retail blue book value. AOPA (The Aircraft Owners and

Pilots Association) will provide this value to members. There are, however, regional market imperfections which may make it reasonable to expect a higher price if you have the time to wait for the right buyer, so you may not want to lock yourself into a blue book value.

An appraisal by an outside appraiser (usually an aircraft dealer) is an option, but appraisals cost money and can be manipulated, so they are not particularly attractive for most co-ownerships. Should you decide on an outside appraisal, make it a requirement that the appraiser be unanimously acceptable to the co-owners.

Co-owners should be required to sign the annual statement of value to signify agreement.

14. Collateralization exclusion.

No co-owner may assign, create a security interest in, or pledge his or her interest in the aircraft without the prior consent of the other co-owner.

This clause states that no co-owner may pledge his or her share of the airplane as collateral for obligations outside the co-ownership. Such a pledge without the written approval of the other co-owners will not stand up in court anyway, but the exclusion clause may short-circuit a messy legal skirmish.

15. Communications. Any official request under the terms of this agreement should be made in writing and delivered to the other co-owner by certified mail or private delivery service.

This section is necessary to provide a traceable avenue of communication in case things go wrong and co-owners need to exercise their rights under the agreement in a contentious situation. Normally all requests can be handled by word of mouth, but the option of an effective, legally binding mechanism of communication must be provided.

16. Arbitration.

In the event that any dispute should arise concerning the co-ownership or the construction of this co-ownership agreement, such dispute shall be submitted to arbitration in accordance with the rules of the American Arbitration Association. The location of such arbitration shall be Anytown, Massachusetts, and the costs of such arbitration shall be born equally by the parties.

All co-ownerships may encounter seemingly irreconcilable differences. Life, however, must go on, and arbitration is the legal profession's well-established way of breaking the impasse. The co-owners agree to submit their dispute to a mutually acceptable lawyer for arbitration and agree to accept the arbitrator's judgment. To ensure impartiality, the lawyer should be required to conduct the arbitration according to the rules of the American Arbitration Association (a standard and well-proven procedure). Arbitration is a good way to keep all but the nastiest disputes out of the courts.

17. Amendments.
This agreement may be amended from time to time by mutual consent of the co-owners.

All agreements are entered into to be amended. As circumstances change it may become necessary to make changes to the co-ownership agreement. This section recognizes the need for making such amendments from time to time and spells out the requirements for making an amendment.

Financing

This is an optional section, applicable only if the co-ownership has incurred any financing of the aircraft. It describes any such financing and spells out the obligations of co-owners with respect to such financing. If the co-ownership has borrowed from a financial institution the following language will usually suffice:

The co-ownership has incurred financing for the aircraft as per exhibit B (the loan agreement). The co-owners agree to abide by the terms and conditions of the loan agreement and will each forward their share of the monthly loan payment to the Treasurer by the specified due date. Each co-owner is individually responsible for any charges incurred by his or her failure to make payments on time. In the event of individual default resulting in foreclosure, the defaulting co-owner agrees to reimburse the other co-owner for all losses incurred by the foreclosure.

If one co-owner has borrowed from another co-owner the following language is appropriate:

The co-owners agree and acknowledge that Mr. Smith's interest in the aircraft has been financed by Mr. Doe pursuant to the terms of a note attached to this agreement as exhibit B. The current principal balance (excluding accrued interest) due on such note is $10,000. Mr. Doe agrees that upon receipt of all future principal payments by Mr. Smith, he will execute an appropriate payments chart appendix modification to this document to reflect the principal payments by Mr. Smith. Mr. Smith acknowledges that his rights in the aircraft and in the aircraft co-ownership are subordinate to the terms of that note and that in the event of loss of the aircraft, death of a co-owner, or dissolution of this co-ownership, the note (together with all accrued interest and principal) shall first be paid before he receives any insurance or sale proceeds. Further, Mr. Smith

*acknowledges that in the event of any uninsured loss or
damage to the aircraft, such uninsured loss or damage shall not
serve to reduce or extinguish the note, which will remain as an
independent financial obligation.*

To sum up, the agreement discussed here is an example intended to
serve as a general guideline for your own agreement. The clauses and
the degree of detail you need depend on such factors as the size of
your co-ownership, the use of the aircraft, and the personal preferences of your group. Be sure to consult a lawyer, and if in doubt, err
on the side of more detail.

Case Study: Piper Arrow III—Two Partners

The aircraft

The airplane is a 1978 Piper Arrow III, considered by many pilots to
be the best Arrow made. The Arrow III has the trapezoid wing similar
to the Warrior II and Archer II and is powered by a 200-hp, normally
aspirated Lycoming IO-360 engine. Subsequently Piper introduced the
Arrow IV, moving the stabilator into a T configuration that was pretty
but degraded pitch control on takeoff and landing. When the production of Arrows was resumed sporadically in subsequent years, the stabilator was quietly moved back down to where it belongs.

Case 5-1. *The Arrow III is arguably the best Arrow made.
(Piper/Montie/Rankin)*

The Arrow III, bought in the mid-1980s, has proved to be a sound investment, steadily increasing in value to a point where it is worth 60% more than the partners paid for it, a return on investment that has consistently beaten inflation over the years.

The Arrow III's plain vanilla handling characteristics make it an ideal complex airplane for relatively low time pilots. If you can handle the Piper Archer, remember to raise and lower the gear (there is an automatic extension mechanism if you forget), and learn to operate the prop control, you can be safe in the Arrow. It is also one of the most economical airplanes in which to maintain complex aircraft proficiency.

Partnership background

There were originally two pilots in this partnership: Don, a 350-hour private pilot, and Jim, a 2000-hour pilot with commercial and instrument ratings. The partners had known each other for a number of years when they both belonged to a flying club. Both wanted the greater flexibility and airplane access provided by personal ownership and decided to join forces in the Arrow III.

Both partners use the airplane for recreational weekend flying, usually on day trips and occasionally overnight throughout New England and upstate New York. Jim also uses it on business during the week in the Northeast. Don's first order of business after buying the airplane was to get his instrument rating in it.

Partnership structure

To maximize liability protection, the partners own the airplane through a subchapter S corporation which they established when they bought the airplane. They have a simple but comprehensive set of by-laws that govern the partnership. While they have gotten along well and have never needed to refer to the agreement, it has proved useful when they took in a third partner a few years after forming the partnership. The third partner didn't work out (he was amicable but was laid back about caring for the airplane to a degree unacceptable to the two original partners); the agreement made it relatively easy to arrange a peaceful parting of the ways.

Finances

The partners share fixed expenses equally and each pay all of own their operating expenses. Both partners maintain separate fuel accounts at their home airport and deal directly with the FBO to buy fuel. The airplane is always left fueled to the tabs. Don maintains the financial records and pays the partnership's bills out of a joint partnership checking account which both partners fund promptly as expenses are incurred.

Early in the partnership Don financed part of Jim's share of the airplane because he didn't want to co-sign a bank note and allow the bank to take out a lien on the airplane. The partner financing was done at arms length, evidenced by a note, with Jim's share pledged as collateral to Don for the life of the loan. The loan has since been paid off as agreed.

The summary expenses of owning and flying the Arrow per partner is currently working out as follows:

ANNUAL FIXED EXPENSES ($)

	Total	(2 Partners) Per Partner
Tiedown/Hangar	1020	510
Insurance	1600	800
State Fees	120	60
Annual	750	375
Maintenance	1500	750
Cost of Capital	4200	2100
Total Fixed Expenses	9190	4595

HOURLY OPERATING EXPENSES ($)

Fuel	16
Oil	0.5
Engine Reserve	7.5
General Maint Res	8
Total Hourly Op. Exp.	32
$/hour @ 100 hr/yr	78

Note that unlike some partnerships, this partnership thinks it is important to account for the cost of capital (currently at a rate of 7% per year). The hourly maintenance reserve has at times proved to be a little low, but the partners have no problem paying in any shortfalls, so the reserve rate has not been increased.

Operations

On the weekends scheduling is on a priority basis. The partners alternate priority weekends. In practice this means first choice for the priority partner and good availability for the other partner unless the airplane is away overnight. Scheduling is often not an issue on the weekends because the partners frequently fly together.

During the week availability is on a first come, first served basis. Since Don rarely uses the airplane during the week, availability for Jim is excellent.

Maintenance

The partners have an unusual arrangement with a local A&P mechanic who is proprietor of his own maintenance facility. The mechanic, who is also a pilot and loves the Arrow, trades maintenance time with the partners for flying time. The mechanic is technically always third in line for the airplane, but in practice the partners are willing to be flexible. The partners enjoy helping out with maintenance and their mechanic allows them to do so.

The partnership experience

Given that this partnership is entering its twelfth year, the experience has obviously been positive. The partners get along well (they frequently fly together), and the partnership gives them a high quality of flying at a comfortably affordable cost.

Aircraft availability is so good that both owners often feel as if they own the airplane on their own. The partners attribute the success of their partnership to their compatibility. They each have a natural tendency to go out of their way to accommodate the other. Once, in a rare midweek scheduling conflict, they even flipped a quarter for the airplane and the loser rented from the FBO.

6

Financing the Partnership

Business people will sometimes tell you that a cardinal rule of business is to do it with someone else's money. Many of us have also taken this point to heart in our private lives. We borrow to buy our houses, cars, furniture, vacations, fancy meals at restaurants, and educations. So why not borrow to buy an airplane? Under the right circumstances, borrowing is a perfectly sound way of gaining access to an airplane you could otherwise not afford. However, for the partnership, bank borrowing poses certain problems that may make it difficult for the group to decide to borrow as one entity.

If the partnership is the borrower, most lenders will want each partner to be responsible for the entire amount of the loan to the partnership. In effect, the lender will want each partner to guarantee that the others will perform. Thus each partner must be satisfied that each of the other partners has the financial capacity to reliably make his or her share of the loan payments. Should a partner fail to pay, the others will be liable to the lender for the entire unpaid amount. Should they not be able to make the payment, the lender will be entitled to repossess the airplane, and the credit ratings of all the partners will be affected because one let the others down.

There are ways for the partnership to get around this problem, but before I get into that it is worth becoming more familiar with the details of aircraft financing.

Fig. 6-1. *Navions have their roots in the P-51.*

Borrowing Basics

Borrowing is really like renting something. You are paying a rental charge (interest) for the use of someone else's money (principal) for a period of time.

You are expected to have a sufficient flow of income to repay the loan in addition to paying obligations you already had when you got the loan. As a fallback source of repayment you are usually expected to pledge to the lender the airplane as collateral (other forms of collateral may be pledged under some circumstances) for the life of the loan. Should you stop making your loan payments, the lender will want to collect the remaining amount you owe by repossessing and selling the collateral.

In addition to paying interest monthly, most lenders will also want you to start repaying principal monthly throughout the period for which you borrowed. They don't want to put faith in your ability to come up with a big lump sum at the end. Also, as you use the airplane (the collateral), its value declines. So, in order to maintain the original ratio of the loan to the airplane's value, the outstanding principal has to decline too.

The problem with borrowing is that it is very expensive. The interest (the rental charge for using the money) comes out of your pocket. Like the rental charge for a car or an airplane at the FBO, it is gone forever. For paying back the principal you at least get to call more and more of the airplane truly your own, but this amount too is cash you have to come up with every month. You must be sure that you can afford to borrow.

Many borrowers have difficulty realistically assessing how much credit they can pay back comfortably and over what period of time. And don't always believe what your friendly banker says you are able to pay. The U.S. financial industry's track record of lending decisions is spotty to say the least. In a strong economy, money is practically forced on borrowers by profit-hungry lenders. When the recession inevitably arrives, hundreds of billions of dollars in loans go into default and are written off.

Ultimately you should analyze the option to borrow just as carefully as you weigh the pros and cons of all the other options that go into making the decisions about your partnership. Following is a discussion of how to figure out what it costs to borrow, and where and how to seek financing.

Borrowing Is Expensive

Lenders are rarely willing to finance 100% of an asset, in this case the airplane. They want to see some of the borrower's own money in the asset financed. Banks are usually willing to lend up to 80% of the fair market value of an airplane to creditworthy borrowers. Fair market value is often determined as the retail blue book value. The examples below consider a typical bank loan for a single-engine, 180-hp, fixed gear, four seat airplane. The fair market value (which should also be your purchase price) of the airplane is $45,000. The bank loan is $36,000, or 80% of the purchase price. The loan is for 10 years at an annual interest rate of 12%. Table 6.1 compares annual costs without any financing and with 80% financing for two, three, and four partners. What do the numbers say?

Cash Expenses

The main point is made by the Loan Payments line. For two partners the additional cash expense is a whopping $2610 more per partner per year with the loan. For four partners the figure is halved, but it is still a hefty extra annual cash expense.

Another way of looking at the extra cost of borrowing is calculating how much you will spend on interest during the life of the loan. The amounts can be surprisingly large. For the $36,000 loan for

Assumptions:
Aircraft Price $45,000; Loan Amount: $36,000; Loan Terms: 10 yrs, 12%/yr; Total Interest Expense: $25,979

($/Pilot)	No Loan	Loan	No Loan	Loan	No Loan	Loan
NUMBER OF PILOTS	2	2	3	3	4	4
$36,000 Loan, per pilot share!	0.00	18,000.00	0.00	12,000.00	0.00	9,000.00
ANNUAL FIXED EXPENSES,						
Non-fin. expenses	2,135.00	2,135.00	1,423.33	1,423.33	1,067.50	1,067.50
Loan Payments	0.00	2,610.00	0.00	1,740.00	0.00	1,305.00
Cost of Capital (non-cash)	1,800.00	360.00	1,200.00	240.00	900.00	180.00
Total Fixed Expenses / yr	3,935.00	5,105.00	2,623.33	3,403.33	1,967.50	2,552.50
Total Op Exp / hr	27.25	27.25	27.25	27.25	27.25	27.25
TOTAL HOURLY EXPENSES						
50 Hours	105.95	129.35	79.72	95.32	66.60	78.30
100 Hours	66.60	78.30	53.48	61.28	46.93	52.78
150 Hours	53.48	61.28	44.74	49.94	40.37	44.27
Interest paid over life of loan	0.00	$12,989.00	0.00	8,660.00	0.00	6,494.00

Table 6.1. *The high cost of borrowing.*

10 years at 12%, the interest is $25,979. This translates to $12,989 per partner in a two-person partnership, $8660 among three partners, and $6494 among four partners. Any way you look at it, that is a lot of money that would have stayed in your pocket.

The bottom line is that in order to minimize the cost of flying it is in your interest to borrow as little as possible. Excessive borrowing may even increase the cost of your flying above the rental option. However, if it takes some borrowing to get you into the airplane you have always wanted, at a cost that beats rentals, and you can make the monthly payments, head for the cheapest source of a loan immediately. Just make sure that your partners can also afford their share of the payments.

Arranging Bank Financing

Most banks think that financing small airplanes flown for fun is more trouble than it is worth. To be done right, it requires that the bank develop a considerable amount of specialized industry knowledge in a market that is fairly small. The chance that someone is going to walk through the door asking for a home improvement loan is far greater than a rush of customers demanding aircraft financing. And home improvement loans are oh so much more comprehensible to bankers than those "dangerous" little airplanes. Fortunately one bank's lack of interest is another bank's niche, and there are a handful of banks nationwide that specialize in financing airplanes. Good sources for finding out about them are aviation trade publications, like *Trade-A-Plane*, where they advertise their services.

The services these banks offer vary, so shop around. They may have different minimum amounts for a loan, different lengths of time for which they are willing to lend, different rates of interest (high, low, fixed, variable), and different maximum percentages of an airplane's value up to which they will make a loan.

Bank lending standards

When a bank lends money for anything, its main concern is how it will get repaid. Some bankers claim that owner pilots are the best credit risk because they will sell the house, the dog, and the kids before missing an airplane payment. A good banker wants the borrower to have reliable sources of cash income as a primary means of repayment, and will want a fallback source should the cash flow dry up. The bank will also want to check if the borrower has a good credit record, a history of paying obligations promptly.

In the case of individual borrowers (as opposed to businesses), typical sources of cash income are regular paychecks, investment income, rental income, royalties, etc. To assess a borrower's ability to repay the loan, a banker must see tangible proof of these sources and amounts and must examine where the money goes, that is, what loans or other liabilities the borrower is already obligated to pay. If, after meeting existing obligations, there is a sufficient amount of cash left over to support the proposed loan, the borrower is in business. Well, almost–as soon as the banker establishes a fallback source of repayment and checks out the borrower's credit history.

The fallback source of repayment is necessary because there is always the chance that the financial statements do not represent the full picture of the borrower's financial condition, and because sources of cash may dry up. You may get laid off. The fallback source is inevitably the asset being financed, the airplane. The banker will file a lien on the airplane with the FAA. This filing establishes the bank's security interest in the airplane and its right to repossess it and sell it to repay the loan should you stop making payments (default).

Financial information requirements

Proof of sources and uses of cash is the financial information you will be asked to submit in the form of income tax returns and signed personal financial statements. For smaller loan amounts, the financial information on the bank's standard application form will suffice. For larger amounts, greater detail may be required, including as many as three years of tax returns. If the partnership is the borrower, the bank will want copies of each partners' tax returns.

The bank's view of partnerships

These requirements seem straightforward for one borrower. But what if a partnership wants to borrow? What information will the bank want, and what obligations does it expect the individual partners to assume? The answer is simple. To the bank, a partnership is the sum of its partners. Thus all partners will have to provide the information required of a single borrower. The bank will determine the partnership's ability to pay by combining the individual partners' ability to pay. In turn the bank will want all partners to be responsible for the loan jointly and severally. In English that means all partners are on the hook for the entire amount. Everybody signs the whole note. The bank doesn't want to be in the business of chasing partners individually. It wants to deal with the partnership as a whole and wants all partners to be financially and morally committed to the partnership's obligations. Let the partners chase each other if necessary.

Recently there have been cases where a bank was willing to lend to only one partner as long as the other partners agreed to the bank holding the airplane as collateral. While this may get a partnership out of the loan, it weakens the position of the nonborrowing partners because if the borrowing partner defaults the bank instantly repossesses the airplane. In all likelihood, the nonborrowing partners would have first shot at buying it or taking over the loan, but if they didn't have the means to do so, the partnership would collapse.

Credit and employment checks

Your credit history will be checked out through the usual credit bureau reports to which all banks subscribe. These reports will also be used to verify the statements you made about your existing debts on your personal financial statement. Most banks will also want to verify your employment.

Loan amount versus aircraft value

Before you approach a bank for a loan it is good to have an idea of the maximum amount you can reasonably expect based on the airplane's value. As stated earlier, the most a bank is typically willing to finance is 80% of the fair market value, usually the retail blue book value. To determine this amount, the bank will want a very detailed description of the airplane, including number of hours on the airframe and engine, avionics, options, and damage history.

If you are not sure of how reasonable the asking price is for the airplane you are considering buying, you can get some good information from your banker indirectly. The price is between you and the seller, so your banker can't explicitly tell you that you are overpaying. But what he or she can tell you (following a detailed description of the airplane) is what the maximum amount is that the bank is willing to finance. That figure will be 80% of the fair market value. You can figure out the rest.

The bank's insurance requirements

Should a bank decide to make a loan to the partnership, it will want to protect its interest in the airplane it is financing, and will therefore have stringent insurance requirements. Check the banks' insurance requirements carefully when you are planning the partnership. These requirements may exceed the partnership's planned levels of coverage, and may mean additional expense, so they are worth knowing in advance. Typical insurance requirements demanded by banks include the following:

- The policy must include the name of the registered owner as primary insured.
- The aircraft must be identified by year, make, model, and FAA registration number.
- Coverage must include all risk of physical damage, in flight or on the ground.
- Coverage should be for the full purchase price.
- Deductibles should not exceed 10% of insured value.
- The policy must name the bank as loss payee.
- The policy must include a breach of warranty endorsement to the bank for the amount financed.
- The policy must contain 30 days cancellation to the bank.

Before a bank disburses a loan, it has to receive proof that the borrower has acquired insurance for the airplane to be financed. Such proof is evidenced by the insurance binder sent by the insurance company to the bank. (For more details on insurance see chapter 7.)

Loan closing and documentation

When the bank approves your loan, it will do a title search (for which you pay a fee) and it will ask you to arrange insurance for the airplane you are about to buy and have the binders sent directly to the bank. When these essential pieces of paperwork are in hand, and all that remains is to actually buy the airplane, you will have to go through a loan closing to sign the loan documents and get the money. You should fully understand these documents, and unless you are a lawyer or an experienced borrower, the information can be overwhelming. These documents are stock forms, and it is an excellent idea to ask the lender for blank examples well in advance of the signing so that you can take your time to fully understand them. Do not hesitate to call your banker or your lawyer with questions as you study the documents.

Documentation of the loan usually consists of three sections: the promissory note, the consumer credit disclosure statement, and the security agreement. All three items may be combined into one document, but often the security agreement is separate (for sample documents, see appendix 3).

The promissory note is evidence of your borrowing. It is the loan agreement. It lists the total amount you promise to pay (principal and interest), the terms of repayment, and all other obligations to which you agree during the life of the loan, such as prompt payment, late charges, events of default, conditions of prepayment, and the like.

The disclosure statement is a statement of the costs and duration of your loan. It lists the amount financed, the total dollar amount of the finance charge, the total dollar amount you will have paid when all scheduled payments have been made, the cost of borrowing expressed as an annual percentage rate (APR), the number of monthly payments and the amount of each payment, and any additional fees and closing costs. It is a statement required by consumer protection laws to ensure that you are not tricked into inadvertently accepting any hidden charges and to ensure that you fully understand the amount of interest you are required to pay. Most agreements have the disclosure statement as an integral part of the promissory note rather than separately restating this information.

The security agreement gives the bank ownership of the airplane up to the financed amount. This is the document that gives the bank the right to file a lien on the airplane. It spells out the conditions under which the bank may take possession and dispose of the airplane,

Fig. 6-2. *You could put this Grumman TR-2 on a credit card.*

and it lays out the obligations of the borrower to insure and maintain the airplane. It also explicitly states the bank's obligation to release its security interest when the loan is paid in full.

Payment problems

Should the partnership not be able to make payments on the loan for whatever reason, it is extremely important to let the lender know sooner rather than later. Prompt notification will result in a much more cooperative banker who may be very helpful in working out a solution which may keep the partnership intact. Whatever the outcome, the banker will be in a good position to preserve your credit rating by an orderly dissolution of the partnership before a serious default. If you are elusive or pigheaded, the bank will repossess the airplane anyway, and nobody will lend you money to bore holes in the sky for a long time to come.

Denied loan applications

There is always the chance that the bank will deny your loan. In this case you have a legal right to receive a written statement of the specific reasons for denial. Treat a denial constructively. Some reasons may be ironed out easily. Perhaps a credit report contained negative information erroneously. Maybe the title on the airplane had a problem that can be taken care of with a few phone calls.

Other reasons may be more difficult to counter. You may already have too much credit for the bank's liking compared to your income. You are paying too much for the airplane and the amount you want financed is too high a percentage of what the bank thinks the airplane is worth. Perhaps a bigger down payment would help if you can afford it, or an additional partner. Whatever the reason for the denial, find out what you need to do to overcome it to get approval the second time around, and never stop trying.

Alternate Sources of Financing

Bank financing is a nice option to have, but it may not suit your partnership's particular needs. Not all partners may want to borrow. Partners may not want to sign for each other. Some partners may not want to disclose financial information about themselves. You may think bank financing is too expensive. Fortunately there are alternatives to the bank.

There are two popular options which keep the partnership free of debt as an entity and have the added benefit of somewhat reducing the cost of borrowing. One is a personal bank loan taken out independent of the partnership and the airplane; the other is financing from one of the partners.

Personal bank loans

The option of a personal loan works best when, for whatever reason, the partners don't want the partnership to borrow. In that case, the partners who need to borrow should consider taking out a personal loan which is granted to them alone, outside the partnership, for general purposes.

The most popular form of such a loan is the home equity loan. If you have a house on which the mortgage is substantially paid down, you have a good chance of borrowing against the house. As a condition, the lender will take a second mortgage on the house, behind the institution to whom you owe the remainder of the original mortgage. An added advantage of this type of loan besides keeping the partnership free of debt, is that the interest is tax deductible, which could mean substantial savings depending on your tax situation. Home equity lines are also likely to be less expensive than other forms of consumer debt, and may be more readily available at fixed rates.

Other options for obtaining personal loans are to take them out against other forms of security which you may be fortunate enough to have available, such as stocks or bonds you do not wish to sell, or long-term deposits you do not wish to withdraw, or the cash value of life insurance.

Partner financing

There may be instances when a partnership as an entity does not want to borrow and the partner in need of financing does not have a personal bank loan available as an option. In this case another partner may be willing to finance the partner in need. Whether or not such an arrangement is desirable is a matter of personal opinion and circumstance. It is, however, fair to say that it generally carries a higher risk of friction than a partnership in which partners do not finance one another.

In spite of the potential for friction, there is something to be gained from financing a partner besides making the partnership possible. The lending partner may be able to realize higher income on a loan to another partner at fair commercial rates than on alternative forms of investment. Think about it. When bank deposits yield 5%, conservative mutual funds yield 10%, and consumer loans go for anywhere from 13% to 19%, a loan at 12% to a partner you know, secured by an airplane with a value far in excess of the loan amount is not such a bad deal.

The key to minimizing the possibility of friction when financing another partner is to make the loan a completely arm's length business transaction, as if it were done between two businessmen who, but for the deal, don't even know each other. The loan should be documented in a promissory note just like a loan from a financial institution, and the lending partner should take a security interest in the airplane and file a lien on it. It is imperative that you get a lawyer to draft the note and security interest. From a legal standpoint this is a very simple procedure and should not require more than an hour or two of the lawyer's time.

If there are more than two partners, a loan from one to another should not affect the rest, except perhaps by forming entangling alliances if decisions do not have to be unanimous. It is true that the lending partner has a lien on the airplane, but by virtue of the partnership agreement that has no effect on the other partners, because their percentage share is recognized in any sale of the airplane. The lien only gives the lending partner a right to the share of the borrowing partner.

7

Partnership Insurance

Insurance is one of the most important considerations of owning an airplane. It protects your investment if your airplane is damaged, destroyed, or otherwise lost. It also protects your other assets from any claims against them arising out of damage or loss caused by your airplane. Insurance is expensive, and, for most pilots, confusing at best. It is not inherently complicated, but it is sufficiently specialized to require careful study to be fully understood.

As with financing an aircraft, the partnership faces certain issues regarding insurance that are of no concern to the individual owner. When the sole owner goes flying he risks only his own investment. He has only himself and his family to answer to in making insurance decisions and living with the consequences. When a partner takes to the air, he risks the investment of all the other partners, as well as his own. A poor insurance choice by the partnership will cause a financial loss to all partners, not only the one doing the flying. Different experience levels among the partners and different preferences in levels of coverage can be additional sources of friction in a partnership.

This chapter outlines insurance basics, walks you through a typical insurance policy, gives you advice on insurance considerations specific to partnerships, and offers advice on where to get insurance.

Insurance Basics

Insurance is protection against the risk of financial loss related to the asset insured. There are two basic elements of aircraft insurance coverage: hull insurance and liability insurance.

Hull insurance, as the name implies, is insurance against damage to the airplane. If the airplane suffers any damage, the insurance company pays to fix it. If the damage results in the total loss of the airplane, the insurance company pays the owners an amount of money equal to the value of the airplane as defined by the insurance policy.

Liability insurance is protection of the airplane's owners against claims by others who have suffered some form of injury or property damage as a result of damage caused by the owners' airplane. This is the big bucks world of the dreaded liability lawsuit, in which everything you own and in some cases even your future salary is fair game as compensation to the victims and their lawyers. A liability suit can ruin you and your family financially, so you must carefully consider the level of coverage you feel is appropriate for your situation.

Other forms of coverage usually included in aircraft owners' insurance policies are some amount of medical coverage for the aircraft's occupants and insurance covering you while you fly an aircraft you don't own. As I walk you through an insurance policy I will examine both forms of insurance and the way to assess appropriate levels of coverage in much greater detail.

Deductibles

An important feature of insurance policies is the deductible. It is the amount you are required to pay when you make an insurance claim. There are different deductibles for in-flight and not-in-flight damage. The amounts depend on the insurance policy, but are usually modest amounts (low to mid hundreds).

Do you need insurance?

Hull insurance is entirely the owners' option, and liability insurance is a legal requirement only in a handful of states. So with annual insurance costs (premiums) commonly running in excess of $1000, who needs insurance? You do! Flying without insurance poses risks of such high financial loss that some form of protection is practically mandatory and doing without is downright stupid. The real question is what type of insurance do you need and how much coverage should you get?

Hull insurance needs are relatively easy to assess, because you can accurately estimate your maximum loss. If you fly without hull insurance, the most you stand to lose is your investment in the airplane. Many pilots claim that they are too good to bend an airplane or, if

Fig. 7-1. *Check into insurance before you buy this Fouga Magister.*

they do, it is their own fault, so they forego or minimize hull insurance coverage. They find out too late that hull damage can easily occur due to circumstances beyond their control, even when the
airplane is under their command. As discussed below, there are various hull insurance options covering air and ground operations at a
variety of prices. Analyze them carefully as you go through the
process of obtaining insurance quotes and decide what option best
suits you. Just bear in mind that if you decide to forego hull insurance
or limit coverage, you are gambling with the amount of your uncovered investment.

The appropriate level of liability insurance coverage is harder
to assess. In a worst-case scenario there is no way to estimate the
maximum amount of a liability award against you. It can run into
the millions. Especially if you have millions. The courts can come
after everything you have, over and above your airplane, and can
even attach future earnings. Thus, as a rule of thumb, it is advisable
to have coverage in excess of your assets. Again, it would be gambling to minimize liability coverage, and if you have a family it
would be totally irresponsible. As you assess your insurance needs,
pay particular attention to the details below regarding liability limit
options before you make your decision. Determine your needs first
and then shop around. Do not compromise on needs for the sake
of a few dollars.

How are insurance rates determined?

The setting of insurance rates is a complex process of blending together a variety of risk factors based on actuarial (statistical) records of events and losses and assigning a price to them. The industry's past loss experience with pilots and equipment fitting your profile plays a key role in establishing your insurance rates. The consistency of rates is compounded by the laws of supply and demand; how many insurers are chasing how many airplanes to be insured. Following a string of high payouts many insurers may decide to exit aviation underwriting. Rates skyrocket. In times of a steadily improving general aviation safety record, evidenced by decreasing payouts, insurers may flock back into the business. Rates plummet. At the end of the 1980s the cost of insuring a Piper Arrow III flown by two relatively experienced private pilots under a fairly typical insurance policy declined in one year from $2250 to $1100 merely because a much larger number of insurers were offering insurance.

Some of the more common determinants of your specific policy costs are as follows:

- experience and qualifications, total flight time, recent flight time, license and ratings held, time in proposed equipment
- type of aircraft
- equipment in aircraft
- home base of aircraft
- how aircraft is stored: hangar or tie-down
- what aircraft is used for: pleasure, pleasure and personal business, commercial use
- how many hours the aircraft is likely to fly during the insured period

Understanding Your Policy

Aviation insurance policies are generally structured in a similar fashion and consist of two components: a standard format of general terms and conditions, and a data sheet spelling out your coverage and terms and conditions specific to your policy. The standard section of the policy provides a set of general definitions to pin down the terms used throughout the policy and defines general terms of coverage for hull and liability insurance in separate sections. Then, in a separate and very important section usually labeled Exclusions, the policy specifically spells out what is not covered. Additional exclusions may also be scattered throughout the policy. There is also a section dealing with general conditions such as renewal, cancellation, and conformity to state laws.

This "boilerplate" policy is customized by a data sheet identifying you as the named insured, identifying the aircraft insured, spelling out the amounts and costs of your coverage, and defining who is authorized to fly the airplane in addition to you and your partners. Any special provisions tailored to your needs are attached to the policy as separate "riders."

The policy language of each insurance underwriter is standard, but there can be substantial differences in language from underwriter to underwriter. Thus it is important to read the policy before signing on the dotted line. Standard policies should be provided for your review.

Let's go through the sections of a standard policy.

General definitions

This section defines the terms used in the policy, such as insured aircraft, accident, bodily injury, property damage, in flight, not in flight, in motion, not in motion, geographic coverage, and the like. From the standpoint of hull insurance it is important for you to understand the definitions of flight and motion. If you intend to fly abroad you may want to pay particular attention to geographic coverage.

Hull insurance

Hull insurance covers your airplane against damage. The decisions you have to make are what types of activity do you want coverage for and what value to insure the hull for. Coverage is available for a variety aircraft operations in flight and on the ground.

- All-risk coverage. The most comprehensive (and most expensive) coverage is "all risk." Under the all-risk option you are covered in flight and on the ground regardless of whether the airplane is tied down, taxiing, or being towed or pushed around.

- All risk not in flight. This is a lower (and less expensive) form of coverage. Under this option your aircraft is covered against damage only on the ground, but regardless of whether or not it is in motion. It is important for you to understand the definition of "in flight." Does it include the takeoff and landing roll? Does it cease the instant the wheels touch the ground? Many self-confident pilots choose this type of hull insurance, reasoning that they can handle a well-maintained airplane once free of the earth's confines. They reason that if they fly safely for long enough, and save the hull premiums, they will be ahead financially as soon as the sum of their saved premiums exceeds the value of the aircraft. Some make it (it can take 20 years), some have to give up flying forever because of an uninsured blown landing or the like.

- Before you and your partners choose this option, bear in mind that in case any one of you is forced down in a forest by a mechanical failure impervious to your impeccable piloting skills, the ensuing damage to the aircraft will not be covered. Only if you and your partners are fully willing to lose your investment in the airplane should your partnership select this option.

- All risk not in motion. This is the lowest form of hull insurance coverage, and covers your airplane only on the ground when not in motion. It provides protection against theft, fire, and vandalism and events such as someone sideswiping the airplane while it is parked. This type of insurance may make sense if for some reason you have a grounded airplane (perhaps awaiting restoration) you are not planning to move during the insurance period.

Regarding the value for which to insure the hull, you may have two choices depending on the insurer: actual cash value and stated value.

- Actual cash value. This is the market value of your airplane, regardless of what it cost you to buy it. This value can be determined by a variety of methods specified by your insurance provider, such as blue book value or an appraisal. The insurance provider will most likely attempt to minimize the value as much as possible. You are not assured of being able to buy a comparable replacement aircraft based on actual cash value.

- Stated value. This option allows you to state a value on which your premium is calculated. It will be the amount paid should you have to collect. The advantage of this option is that you can establish a replacement value for your airplane, better protect the partnership's investment, and get back in the air with the least amount of financial disruption.

Beware of the pitfalls of overinsuring. If you overinsure, the insurance provider may find it less expensive to repair the aircraft instead of writing it off and handing you over enough money to buy a new one. At any rate, the insurance company won't allow you to overinsure beyond a certain point.

Deductibles vary depending on your preference and the phase of flight in which the damage occurred. A willingness to accept a higher deductible may result in lower premiums.

The types and amount of hull insurance and the deductibles specific to your policy can usually be found on your insurance policy's data sheet.

Liability insurance

Liability insurance covers you against claims by persons to whom the operation of your airplane caused some form loss or damage. They could be your passengers or anyone outside the airplane or their heirs and survivors. Loss or damage is separated into two categories: bodily injury and property damage. Your policy sets limits to liability coverage, with which you should be thoroughly familiar. The limits offered by different insurers may also differ significantly. The structure of liability coverage limits should be one of your top priorities when you shop around for insurance.

Most policies today have combined liability coverage and limits for bodily injury and property damage, but some may not, so be sure to check. Let's take a closer look at typical types of liability limits.

- Each occurrence. This amount is the maximum amount of liability insurance coverage you will receive for each accident or incident. It is tempting to choose the lowest amount of coverage available to save on insurance premiums. The fact is that these amounts are unlikely to be sufficient in case of any major mishap. Weigh carefully the cost of extra coverage, and don't be too tightfisted. In this era of multimillion dollar liability awards, liability coverage of at least $1 million is considered prudent.

- Each person. This is a sublimit of each occurrence. It is the maximum amount that will be paid to each claimant per occurrence. Here is where a spectacular amount of total liability insurance can quickly erode. Many policies will set very low limits per person. A total liability limit of $1 million is worthless when your runaway prop drilled an irreparable hole in a sole provider to a family of eight and all they will get under the policy is $100,000 because that is your per person limit. Be sure that the per person coverage is substantial. The best policies are the ones with no per person restrictions.

- Also the per person limit is usually the maximum amount for which medical expenses will be paid to each occupant of your aircraft. Some policies spell out this medical coverage in a separate section.

- Family member limits. This is where many insurance companies provide practically negligible amounts of coverage. Limits on spouses of 25% of the per person limit and on children of 12.5% of the per person limit are common. The theory is that immediate family members are not likely to sue each other, nor should they be tempted to.

Exclusions

What the large print giveth, the small print shall taketh away. Beware of exclusions. Read them carefully and make it a top priority to understand them fully. Many exclusions are standard, such as the ones prohibiting commercial operations or trafficking in narcotics. But others may slip in that can have a much greater impact on your coverage. Let's take a look at some typical exclusions.

FAR violations Some policies have an FAR violation exclusion, meaning that if you violated any FARs you are not covered. This is bad news, because it is practically impossible to be involved in a major mishap without some FAR violations. Furthermore, insurance companies have used this exclusion to deny claims when the violation of the FAR had nothing to do with the cause of the accident. An example is a denied claim because the proper aircraft documents were not on board when the airplane struck a landing light and folded a gear on landing. A gross violation will always be challenged, but a FAR exclusion may be a sign of an intent by the insurance provider to stiff you in case of trouble.

Flights on legal airworthiness certificate waivers An exclusion of flights on an airworthiness certificate waiver excludes some perfectly routine operations, such as flights on a ferry permit. This is not a major problem, but be aware of it and be sure to get additional coverage if the need arises.

Pilot qualifications This is an important exclusion that deals with pilots other than the owners flying the airplane. The exclusion will usually refer to the data sheet which will spell out the requirements nonowner pilots must meet. The most restrictive option requires that only pilots named in the policy can fly the airplane. The less restrictive alternative imposes conditions to be met by nonowner pilots in order to qualify to fly the airplane.

Some policies may even exclude maintenance personnel conducting test flights in connection with performed maintenance. Be sure to check your policy because most maintenance personnel simply assume they can test fly the airplane.

Fig. 7-2. *You'll need to be tailwheel-qualified to solo this Chipmunk.*

Instruction exclusions Many policies exclude the use of the aircraft for instructing nonowners. So if you are an instructor and want to give an occasional lesson or check ride to a friend (out of the goodness of your heart, because accepting payment would be a commercial operation), be sure to check the exclusions.

In-flight operation by a nonowner This exclusion states that nobody other than the people authorized by your policy can operate your airplane in flight, even if you are in it. If your policy has such an exclusion, letting a friend handle the controls in flight invalidates your policy. How will they know, you ask? Maybe they won't, but sometimes they find out.

Off-airport landings Some policies specifically exclude intentional off-airport landings. Pilots of Piper Cubs heading for a surprise visit to Uncle Harry's ranch take note.

Normal wear and tear or mechanical failure This exclusion states that insurance will not pay for repairs required as a result of normal operations. The insurance provider will not buy you a new engine when the old one wears out. This is clear enough, but the mechanical failure exclusion can get murky. It basically means that if you have an in-flight mechanical failure and no damage beyond this mechanical failure occurs, the insurance company will not pay. However, if as a result of the mechanical failure other damage was sustained (say, during a forced landing), the insurance company will pay for the damage but not the original mechanical failure. This can get especially confusing when an engine failure leads to a forced landing and the airplane is totaled. Will the insurance company fully pay out? Will it reduce payment by the estimated amount of the mechanical failure? Check with your insurer.

Military airport waivers Military airports require civilian users to sign liability waivers. Therefore some insurance policies exclude the use of such airports. Others state that in spite of your signing the waiver, your insurance policy remains in force.

War, revolution, insurrection This exclusion states that if you are caught in such activities don't look for an insurance check when your airplane blows up. This is a standard insurance industry exclusion and also applies to many other forms of insurance.

There may be many other exclusions, depending on your insurer and the coverage sought. Let the buyer beware.

Partnership Insurance Considerations

Now let's take a closer look at the most common insurance issues specific to aircraft partnerships and how to deal with them.

Partnership hull insurance

The choice faced by the partnership is what level of coverage to get. Should it be all risk, all risk in motion, or all risk not in motion? As stated earlier, there are pilots who are so confident of their flying skills that they get only all risk not in flight hull insurance. If a partnership selects all risk not in flight coverage or less, it implies that each partner has complete confidence not only in his or her own flying abilities but also those of all the other partners. It is one thing to think that you fully trust the skills of your partners when you are writing out the insurance payment check and quite another to say "Aw, forget about it" when a partner has just wiped out your $20,000 investment because he forgot to put the gear down and the partnership did not have all-risk insurance.

Experience has shown that the fingers start pointing quickly at the partner in command in case of trouble, so all-risk hull insurance is strongly recommended. The good news is that the incremental increase in insurance costs for all-risk coverage is spread among the partners and should not amount to an especially large additional financial burden per partner.

The selection of actual cash value or stated value coverage is a matter of preference, as it is for the sole owner. Just remember that stated value will give you a known amount (less deductibles); be sure that all partners are satisfied with the choice.

Who pays the deductible?

This question is simple on the face of it. If a partner was in command at the time of the accident or incident he or she should logically pay the deductible. If the aircraft was not in motion it is logical for all partners to pay equal shares of the deductible.

Some partnerships consider an aircraft away from home base to be under the control of the partner using it and require that partner to pay the deductible if an away from home accident or incident occurs, even if the aircraft was not in motion at the time.

Partnership liability insurance

It is the selection of the appropriate amount of liability insurance coverage that poses the greatest insurance challenge for partnerships. The reason is simple. Different partners have different ideas of how much liability protection they need. A collective decision becomes particularly difficult if there is a large difference in the amount of assets the various partners have. A partner just out of school with few assets but a good income may be paired with a partner who has been working for ages and is worth a very substantial amount but does not want the hassle of sole ownership. While $500,000 in liability coverage may be sufficient for the partner just out of school, the one with substantial assets may want at least $2 million of coverage. The partner who wants higher overall coverage may also want significantly higher per person per occurrence coverage.

The first step to sorting out levels of partnership liability coverage is for the partners to see if a coverage amount sufficient for everyone can be obtained at an acceptable price. Depending on the number of partners, the additional cost to each partner of substantially higher liability coverage may be quite low.

If the partners can't resolve their differences regarding levels of liability coverage, the partners seeking higher coverage can purchase additional liability insurance independent of the partnership. Ask your insurance provider for assistance.

A pro rata division of liability insurance costs with the partners seeking higher coverage paying proportionally more is not an equitable option. All partners would benefit from the same level of high coverage when they fly, regardless of their share of the payment, therefore partners paying proportionally more would be subsidizing the partners paying less.

Liability implications of the legal form of the partnership

A big concern in partnerships is the exposure to liability claims of all partners caused by the actions of one partner. The problem can be especially disastrous if the partner causing the liability claim did something that invalidated the partnership's liability insurance. Here's where incorporation can pay off. Work with your attorney to determine the ownership structure that makes the most sense for you.

Insurance cost implications of different experience levels

Generally insurance rates will be higher for pilots with lower levels of experience. In the case of an aircraft flown by a number of pilots, an insurance company will determine the insurance rate based on the experience level of the least experienced pilot. In partnerships where there is a wide range of experience, this means that if insurance expenses are shared equally, partners with higher levels of experience are subsidizing the partners with less experience.

If this subsidy is substantial the partners may wish to divide insurance payment shares more equitably. The way to do it is to get separate insurance quotes. In addition to the actual insurance quote, a hypothetical quote should be obtained assuming that only the experienced partners would fly the airplane. The payment share of the experienced pilots should then be determined by an equal distribution of the hypothetical quote. The difference between the actual quote and the hypothetical quote should be shared by the less experienced partners only.

Consider, for example, a partnership of three pilots in a Warrior. Two are VFR private pilots with 150 hours each. One holds a commercial license with an instrument rating and has 3000 hours. The actual annual insurance quote is $1200. This would imply that each partner's share is $400. However, if all three pilots had a commercial license with an instrument rating and 3000 hours of flying time, the insurance premium would be $800. This amount divided three ways is $267. Any amount in excess of $267 paid by the more experienced pilot in the actual partnership would be more than required from a pilot with his level of experience, and therefore is a subsidy of the less experienced pilots. Thus the fair shares of the annual insurance payment would be $267 by the experienced pilot and $467 each by the two less experienced pilots ($400 plus half of $133, the extra amount to be paid because of their lower level of experience).

Prorating insurance payments for amount of time flown

Some partnerships prorate insurance payments on an hourly basis. The partners who fly more pay more of the annual insurance costs. This arrangement is most often used when there is a big difference in the amount of time the various partners fly the airplane. The validity of this rationale is a matter of opinion. On the one hand, the partners who fly the airplane more expose it to risk more often. On the other hand, all partners are free to fly the airplane as much or as little as they care to, yet, regardless of anticipated annual flying time, the insurance expense for the year would be approximately the same and would have to be paid anyway. In my opinion, insurance is practically a fixed annual cost of getting airborne, so, except for big differences in the partners' flight experience levels as addressed above, I favor an equal sharing of insurance expenses.

The importance of a good co-ownership agreement

As you can see, partnership insurance arrangements can be quite involved and the consequences of a bad insurance decision may be substantial. All partners should clearly understand and fully agree to the insurance arrangements entered into by the partnership. It is therefore imperative to spell out in writing the way the partnership handles insurance, and the place to do this is in the partnership agreement with the advice of a lawyer.

Insurance Sources

Within the insurance industry aviation insurance is a specialized branch offered by specialist firms. Generalist agents who insure your house and car could possibly get you a quote too, but they will have little understanding of either the industry or your needs, so go to the specialists. Aviation insurance is generally available from two sources: underwriters and insurance brokers.

Underwriters

Underwriters are the firms that actually take the risk on you when you get insurance for your airplane. They offer insurance directly, cutting out the middleman. Some of the middleman's commission they pass on to you, most of it they keep. The benefit of insuring directly with an underwriter is some savings on the cost of your coverage. The drawback is limited flexibility. Underwriters tend to offer stock policies with little scope for tailoring them to your specific needs. They are not in the business of advising you regarding alternative options in the insurance industry. You are expected to know what you want and they will give you a quote. Another disadvantage is that you have to do the donkey work of contacting a variety of underwriters for competing quotes. This is time consuming and you may not be aware of all the available underwriters due to your lack of experience in the industry.

Insurance brokers

Insurance brokers are middlemen. They help you assess your needs and go out to the underwriters to get the coverage you want at the best price available. An insurance broker expects to spend a considerable amount of time with you assessing your needs, explaining to you the various options available, and tailoring a policy to best meet your requirements. A good broker is perhaps the best option for personalized service. Brokers are paid by commission from the underwriter for policies placed. The commission is usually about 10% to 15%. A good broker may be able to save you that amount or more on the policy price by shopping around or getting discounts unavailable to you directly.

Competing broker quotes

You should be aware of a peculiarity of obtaining insurance quotes through a broker. Once you appoint a broker to get you a specific quote, it is industry practice to consider that broker to have an exclusive right to obtain quotes for you for a reasonable period of time. Thus, if you appoint a broker and then call an underwriter directly to shop around after your broker has had contact with the underwriter, you will be told that a broker is already handling the case and you will not be given a quote. This practice protects a broker from having clients "stolen" by competing brokers. If you don't like what your first broker is telling you, the proper way to get competing quotes is to go to other brokers consecutively.

Choosing an insurer

With the limited insurance industry experience most of us have, choosing an underwriter or broker can feel like going to a pot luck dinner. There are, however, some simple ways to take some of the uncertainty out of choosing an insurer. For best results make an effort to apply as many of these techniques as possible.

Talk to the insurer   There is no reason not to interview the insurer, be it an underwriter or a broker. Find out how long the insurer has been in business. Ask what professional aviation background its agents have. Are any of them pilots? Underwriters may be public companies which publish annual reports you can get. You may not have the qualifications to analyze the financials in the annual report, but it should be full of a wealth of other information that will give you some insight into the company. Chat with a broker about his or her background, training, aviation interests and qualifications, and ask for references.

Word of mouth   This is an excellent source of information if you do a good job. Talk to as many airplane owners and FBOs in your aviation community as you can. Ask them about rates and quality of service. Make every effort to talk to airplane owners who have had to collect under a policy. Were they paid what they expected? Were they given a hard time? Did the insurer try to pressure them into a settlement they considered unfair? Did the policy cover what they were led to believe or were there surprises regarding coverage?

Aviation associations   From time to time aviation associations such as AOPA and EAA (Experimental Aircraft Association) enter into agreements with insurers to provide coverage for association members. While the associations are unlikely to explicitly endorse the insurer to you personally, you can take some comfort in the link between the two institutions and may qualify for the programs being offered.

Banks and other financial institutions   A good source of information about insurance providers are banks and other financial institutions who lend for aircraft. They require that the aircraft they finance be insured and are familiar with the industry.

Regulatory and rating agencies   Call your state insurance commission to see if there have been any complaints against the insurer you are considering. There are also several independent insurance rating agencies which in essence give grades to the insurers they cover. The A. M. Best Co. is the most widely used. Get an understanding of the grading system and see how your insurer stacks up. Sometimes the rating agencies' books are hard to decipher. Get your insurance agent to help.

Beware of Multiple Insurance

Every once in a while someone gets the bright idea of buying two separate insurance policies for one aircraft. More often two policies end up being in place for one aircraft when the owners want to change insurers and buy the second policy without simultaneously canceling the first one. *At any time when two different policies are written for the same aircraft both policies are invalidated and you will be without coverage.* So, if you change insurers, be careful how you go about it.

Insurance Worksheet

See appendix 3 for an insurance worksheet designed to help you collect data on insurance options. It is a helpful summary list of the issues to be addressed as you talk to your broker or underwriter.

Case Study: Pitts S-2B– Two Partners

The aircraft

The aircraft is a 1984 Pitts S-2B, a two-seat aerobatic aircraft. It had two previous owners before the current partnership bought it. It is an ideal aircraft type for a partnership because when used for its special purpose, aerobatics, it is seldom flown for more than 45 minutes a sortie, so scheduling is rarely an issue.

Partnership background

This is a partnership of two aerobatic pilots, Ford and John. The airplane is the first one Ford has owned. John has owned several other aircraft before the Pitts. Ford and John both learned aerobatics at about the same time from the same instructor and also knew each other from work. Both wanted to fly a high-performance aerobatic

Case 7-1. *A Pitts S-2B in partnership can turn individual dreams into reality.*

aircraft but neither could afford one on his own, nor did they see any sense in the sole ownership of an airplane that is flown on one or two 30- to 45-minute sorties a day per pilot, if used for its intended purpose.

Ford and John talked about forming a partnership for about a year before taking the plunge. One early hitch that held up progress for a long time was the question of where to base the airplane. Each partner wanted it at an airport 15 minutes from his own house, requiring the other partner to drive about an hour to get to it. Ultimately Ford decided that he was in a better position to make the long drive than John and they started looking for a Pitts in earnest.

They chose a Pitts S-2B because they wanted a two seater (Ford particularly enjoys sharing the aerobatic experience with friends) that is also competitive through the top levels of aerobatic competition. The S-2B offered the most performance for the dollar among high-performance aerobatic two seaters.

They looked for the airplane primarily by word of mouth within the tightly knit aerobatic community. After checking out a few airplanes they found the one they bought within 120 miles of their chosen home field. To evaluate the airplane they relied on the assistance of two pilots highly experienced in flying and maintaining the Pitts. Their instructor, a U.S. National Champion who had competed in a Pitts for years before moving into a monoplane, checked out the airplane from the pilot's perspective. Another friend, who had built several Pitts and would do the maintenance on the one the partnership would acquire, did the prepurchase inspection. The partnership has been in existence for about year and a half.

Partnership structure

Ford and John own the Pitts as co-owners, evidenced by checking off "co-ownership" on the form of ownership on the FAA registration. The co-ownership is not incorporated, and Ford and John do not have a written partnership agreement. Like many pilots who enter into a partnership for the first time, they feel that they can make the partnership work on trust and are ambivalent about having written rules. On the one hand they feel that a written agreement would depersonalize their sense of ownership, but on the other hand they recognize that having a written agreement on some fundamental issues, such as succession in case of the death of a partner, may be useful.

While they did not codify in writing their understanding of how the partnership is structured, Ford and John discussed all the issues at length and have a clear verbal understanding of the partnership's rules. So far their personalities have made the partnership work, but should any unforeseen serious dispute arise, a simple written co-ownership agreement could prove beneficial. Ford feels that while an unwritten understanding such as theirs may work for a two-person partnership if the personal chemistry is right, it would be risky to form a partnership of three partners or more without a written agreement.

Regarding insurance, both partners wanted full hull coverage. Their level of liability coverage was set by the fact that they wanted to enter aerobatic competitions, which requires a level of liability protection specified by the contest organization.

Finances

Ford and John paid equal shares to buy the Pitts. They share the fixed expenses equally, and each pay the operating expenses in full for the time they fly the airplane.

Both partners borrowed some of the purchase price of the airplane. One partner found sources not linked to the airplane and one took out an aircraft loan. The other partner had to co-sign the collateral agreement for the aircraft loan, which entitles the bank to repossess the airplane if the borrowing partner defaults.

In practice the other partner can keep the bank from repossessing the airplane by paying out the loan if the borrowing partner defaults (and ending up owning the entire airplane) or by making the monthly payments to prevent a default. If the latter course is taken one partner is financing the other partner, and this needs to be documented in writing. Stepping in to make the monthly payments should only be a temporary measure to keep the bank at bay until the airplane is sold and everyone is paid out, or until a new partner is found.

The expenses of the Pitts S-2B work out as follows:

ANNUAL FIXED EXPENSES ($)		(2 partners)
	Total	Per Partner
Tiedown/Hangar	2,520	1,260
Insurance	3,100	1,550
State Fees	75	38
Annual	750	375
Parachute Packing	300	150
Total Fixed Expenses	6,745	3,373

HOURLY OPERATING EXPENSES ($)	
Fuel	41
Oil, etc.	3
Engine Reserve	22
Propeller Reserve	2
General Maint Res	10
Total Hourly Op. Exp.	78
$/hour @ 100 hr/yr	112

Ford and John found that when they were making their projections they estimated fixed expenses accurately, but were off on the operating expenses, particularly the cost of ongoing maintenance, which they underestimated.

Operations

The partners have a single joint checking account for the airplane. They each pay their share of the fixed expenses into the account once a month. They keep a flight log in the airplane that records the date of flight, route of flight, tachometer time, fuel totalizer reading, and the fuel added after the flight. The partners have agreed to an hourly operating expense rate per flying hour, which includes fuel (a "wet" rate). They calculate the hourly operating expense based on this wet rate after each flight and pay it into the airplane account.

They maintain one fuel account with their home base FBO and pay the bill monthly from the airplane account.

If a partner buys fuel away from home base, he pays for it out of his own pocket, pays the hourly "wet" rate into the airplane account, and reimburses himself out of the airplane account for the fuel expense paid away from home. The partners have decided that variations in fuel prices away from home are not sufficiently significant to merit readjusting the wet rate (which is based on the home base fuel price) on a case-by-case basis.

Scheduling is informal because each aerobatic flight is so short. As Ford puts it, "If you go to the airport and the airplane is gone you know that it will be back in less than an hour." When a partner wants to take the airplane for a day or overnight, he calls the other partner to ask. The one scheduling request that can put the airplane on the road for more than a day or two and automatically takes precedence over all other uses is participation in an aerobatic contest by either partner.

The airplane is hangared both at its home base and when it is away from home.

Maintenance

The Pitts is maintained by a friend of the two partners who is an A&P mechanic and has built several Pitts Specials. One of the two partners was more willing to do FAA authorized owner maintenance on the airplane than the other, who was uneasy with the idea given the airplane's use. As a result, they have deferred all work to their A&P friend, whom they assist occasionally.

Both partners are committed to being conservative on getting maintenance items done. They have found that when they choose to repair an item instead of replacing it, they usually end up having to replace it anyway. They have full confidence in their mechanic, and whatever he says needs doing gets done. Any discretionary maintenance or equipment upgrade decisions are by consensus.

Maintenance is reserved for per flight hour. The hourly maintenance reserve has two components: an engine overhaul reserve and a general reserve for ongoing maintenance.

Ford, who likes working with spreadsheets, keeps the records. He says that the maintenance reserve built up very rapidly because of the hefty engine reserve component. This reserve is lumped into the single account maintained for the airplane rather than being kept in a separate account. As a large surplus sum accumulated in the account in relation to day-to-day expenses, both partners acquired a tendency not to send in their hourly payments every time they flew. Right now when an arrears of a few months builds up, Ford updates the books and both partners pay their share of the outstanding payments.

Both partners have a similar outlook on the cosmetic care of the airplane. It gets wiped down after every flight, making airplane washing details unnecessary.

The partnership experience

The partnership experience has been absolutely positive. "We both recognize that neither of us could be doing this type of flying without the other," says Ford.

Both partners are picky about certain things and there have been some frank but friendly exchanges about differences of opinion, but at the end of the day they are both sufficiently flexible to sort out the issues. Points of discussion have been about owner maintenance, an occasional scheduling snag for full-day or overnight use, and when to temporarily ground the airplane in the winter to protect it from the cold and slush. Whenever these issues have arisen, they've been amicably resolved, because in the end both partners are very accommodating, but neither hesitates to speak up if something is on his mind.

8

Operations and Maintenance

Operating the partnership is largely a matter of implementing the partnership agreement (which is the basis for all the operational elements) and maintaining a good flow of information among the partners. The extent to which you and your partners keep records will depend on the partnership's size, its level of activity and the personal inclination of the partners to be meticulous and well informed. It is fine to get by with minimal record keeping, as long as all partners recognize the limitations of such an approach and nobody pretends to be surprised when the muddled scribbling on the backs of envelopes causes great confusion and disagreement.

Scheduling, keeping track of hours flown, recording and resolving maintenance squawks, and monitoring aircraft performance are the key operational elements of an aircraft partnership.

Scheduling

Many partnerships do the scheduling by word of mouth, but if there are more than two partners it is a good idea to take turns being the scheduling coordinator and keeping some form of written schedule. The schedule you design will depend on the arrangements you and your partners have for using the airplane.

Many partnerships have some priority system where one partner has the airplane for a day, a weekend, or a whole week, and the other partners have access to the airplane only if the partner with priority decides not to fly. Such a system is usually on a set rotation and

Fig. 8-1. *Your operations procedures should address operating on the grass.*

can therefore be published in advance and sent to all the partners. The scheduling partner is then responsible for coordinating the use of the aircraft over and above the needs of the priority pilot.

It is the responsibility of the pilot with priority to let the scheduler know when and if the airplane will be available during their priority period. The scheduler can then allocate the airplane among the nonpriority partners on a first come, first served basis. For partners without priority, and if there are periods (say, during the week) when nobody has priority, it is advisable to block out some comfortable window of time that will allow leisurely use of the airplane, making it available for others on a reasonable time frame. Dividing the day for nonpriority partners between morning use and afternoon use usually works quite well.

It is a good idea to provide the scheduler with a scheduling calendar used exclusively for the airplane to avoid confusion and to maintain a booking record. If your partnership is on a priority system, the priority information should be entered in the calendar in advance. It is also strongly recommended that the scheduler have an answering machine and be committed to returning phone calls within a specified period of time. An e-mail system is even better if all have access. There are few things more frustrating in an aircraft partnership than an unreachable scheduler.

If your partnership does not have a priority system, it is still advisable to agree on standard time slots of generous duration, depending on the type of flying you and your partners generally do. If your flying is mainly local or instructional, half-day limits are convenient. If you all take day trips frequently, full-day availability should be your standard. If there is frequent demand for the airplane for longer periods of time you should be on some form of priority system instead of an ad hoc arrangement.

Trial and error will prove what works best for your partnership, but it is important to have a good scheduling system to keep everyone happy and maximize the use of the airplane.

Flight Log

Keeping an accurate log (Table 8.1) of when the airplane was used and by whom is indispensable for an effective billing system. The best solution is a flight log kept in the airplane to be filled out by the pilot immediately after every flight. This log can be used to record a variety of information, but at a minimum it should include the following:

- date of flight
- route of flight
- pilot's name
- beginning tachometer time
- ending tachometer time
- total duration of flight
- remarks

This log should be the basis for charging partners for the hourly use of the airplane. It should be proofed frequently against payments received by the partner keeping the financial records. It should be standard procedure for partners to know the hourly rate to be paid for the airplane and to send the partner handling the financials a check for the appropriate amount immediately after each flight (with the appropriate notation, i.e., N37RW, 1.7 hours, $58.00).

Invoice and Check Box

Some partners are notorious for slovenly administrative habits, including being chronically late sending in the checks they owe. Others are frugal and hate to waste a postage stamp every time they have to make a payment for a few hours of flying. For these people

DATE	PILOT	TACH	FLT TIME	ROUTE

Table 8.1. *Aircraft log.*

there is a convenient solution in the form of an invoice strongbox. It is a metal box available in most stationery stores. Its top surface contains blank invoice forms. It has a slot on the side and room for lots of invoices and checks inside, and it can be kept locked. The partnership can get such a box into which every partner can slip his or her checks along with a completed invoice immediately after a flight.

The partner handling the finances can clear out the box periodically (say, once every 2 weeks). Nobody will have an excuse for not making a payment, and no postage will be wasted. Some people dislike the idea of leaving uncashed checks in the airplane, but if the box is kept away from prying eyes, no cash is put in it, and it is frequently emptied, the convenience may be worth the minimal risk of having to stop a few checks.

Maintenance Squawks and Maintenance

It is important for partnerships to keep a record of maintenance squawks on some standard form (Table 8.2). The sole owner knows if there is a problem and whether or not he or she has had it fixed, making a squawk record an unnecessary burden. But for partners who may not always see each other and don't keep close tabs on each other's flying, a record of mechanical squawks and their resolution can be invaluable.

Bear in mind that a good squawk sheet has a dated column not only for noting the problem, but another one for noting the resolution. Any mechanical glitch that makes the aircraft unairworthy should be considered an emergency to be immediately reported to all partners in addition to being entered on the squawk sheet. A note declaring the aircraft unairworthy should be prominently displayed in the cockpit.

Any squawks should immediately be brought to the attention of the partner responsible for making maintenance arrangements, who should get them fixed in the most timely fashion. From an operations perspective, maintenance should be handled as agreed by the partners

AIRCRAFT SQUAWK REPORT		N:	Tach:
Date Noted:	Pilot:	Date Resolved:	By:
Squawk:		Resolution:	
Signature:		Signature:	

Table 8.2. *Maintenance squawk sheet.*

per the terms of the partnership agreement (see chapter 5, "The Partnership Agreement").

Aircraft Performance Records

Professional flyers have long made it a practice to keep records of in-flight aircraft and engine performance parameters. For the pilots of small airplanes on pleasure flights this is an option, but a useful one. It will teach users a lot about the performance of their airplanes, it can provide early warning of mechanical problems, and it can clear up uncertainties about performance by providing a historic reference. Items most commonly logged are power settings, fuel consumption, engine pressures and temperatures, altitude, outside air temperature, and indicated airspeed (see Table 8.3).

Inspections Due Date Chart

A very useful chart, prominently displayed in the cockpit for best effect or available with the flight logs, is a record of the due dates

Fig. 8-2. *The Turbine Bonanza is an interesting aftermarket modification. (Courtesy of Tradewind Turbines.)*

TACH END:	DATE:		N:			PILOT:		
	ROUTE:							
(-) TACH START:	OAT	ALTITUDE	IAS	MP		RPM	MIXTURE	
= FLIGHT TIME:	OIL PRESS	OIL TEMP	EGT	CHT		GAL/HR	TAS	
REMARKS								

Table 8.3. *Aircraft performance records form.*

of the next required inspections, such as the annual, transponder and altimeter checks, and VOR checks as applicable (Table 8.4). It is important for someone to take charge of keeping the chart current, but if a recent inspection is inadvertently not recorded, the error, while annoying, is on the conservative side and will not compromise flying safety.

Flight Planning and Briefing Forms

A partnership may find it convenient to have an indexed binder of preprinted blank flight planning forms, weight and balance forms, weather briefing forms, and flight plans. It may even be worth keeping the used forms in a binder for reference and as a record of the airplane's activities.

Financial Record Keeping

Keeping the financial records, collecting payments due from partners, and making payments on behalf of the group is one of the more thankless tasks of an airplane partnership. It requires a lot of administrative discipline, it can be somewhat time consuming, and if it gets messed up it can be a big source of friction among the partners and may even cause the partnership to break up.

RECURRENT CHECKS DUE	N:		
Enter date or Tach hours when next check due. Cross out superseded date..			
ANNUAL INSPECTION:			
100 HOUR INSPECTION:			
TRANSPONDER CHECK:			
STATIC/ALTIMETER CHECK:			
VOR CHECKS:			

Table 8.4. *Inspections due date chart (to be kept in the aircraft).*

The partner "volunteered" to handle the financials should have the requisite skills and the discipline, and should be aided by the other partners in every way possible. Think long and hard before you make the financial record-keeping task a rotating responsibility among partners. If one partner is especially good at it, you may all be better off by allowing him or her to permanently handle the financials and find other ways for the rest of the partners to make a contribution.

The basic task

The financial management of the partnership consists of the following basic responsibilities:

- collecting from the partners in a timely fashion the money they owe the partnership.
- paying the partnership's bills in a timely fashion.
- keeping accurate records of the partnership's financial transactions.
- providing an accurate periodic accounting to all partners of the partnership's financial transactions.

A big part of the job is keeping after the partners to send in their share of expenses. It is not advisable to cover for a partner and pay bills prior to receiving each partner's share of the payment. Practically all bills are presented with sufficient lead-time to allow all partners to forward their share of the money before the bill needs to be paid. If a partner lacks the discipline to make timely payments, an arrangement should be made with him to put an advance into the partnership for upcoming bills.

Checkbook accounting and partnership accounts

In spite of the dreary administrative responsibility, keeping the partnership's books is a fairly simple financial task. It is best tracked on a cash basis. This is a record of the movements of cash in and cash out of the partnership. In fact, accounting for the partnership is like keeping a checkbook. The partners' payments into the partnership are the deposits. The payments are the expenses for which the checks are written.

The tool for a good partnership's financial transactions is the partnership checking account. It is usually opened in the name of the partnership. All partners should have signing power on the account, and withdrawals above a certain amount should require the signature of two partners. Many easygoing small partnerships open a partnership checking account and simply use the checkbook as the record

Fig. 8-3. *The Champ has low maintenance costs if you check it out carefully before buying it.*

of financial transactions. This is certainly a viable form of minimum financial record keeping from which periodic summary reports of the partnership's financial activities and condition can be extracted.

The checkbook is the original financial record of the partnership account, and more detailed financial reports must reconcile to it. For partners interested in more closely tracking partnership financial transactions and their own share of it, a more detailed yet still simple form of record keeping may be a worthwhile effort. If it is diligently kept current it will provide accurate up-to-the-minute financial information broken down by partner.

Financial records and software

Today one of the easiest ways to keep track of partnership expenses and income and generating reports for the partners is by using some form of electronic spreadsheet or easy-to-use accounting software, such as Quicken. Alternatively, for a small partnership the old-fashioned manual system works equally well.

(See Fig. A2-3. in Appendix 2.) It is a financial reporting sheet for a partnership of three pilots. The income side is simply a breakdown by partner of the deposits made into the checking account for all the expenses of the partnership such as tie- down, insurance, loan payments, as well as a payment per flying hour which includes the engine and airframe reserve (but does not include fuel if in your partnership each partner pays fuel individually and leaves the airplane topped off after every flight).

The expense side is the record of payments transposed from the checking account. Ideally each partner would pay his share of each expense item with a separate check which would make things really clear. In practice, however, partners may send one combined check to cover a variety of expenses that are due at the same time. If they do so, they should include a note breaking down the total payment into its constituent parts. The partner keeping the books can then enter in the description column an accurate dollar breakdown of what each check was for.

On a periodic basis, and at any rate annually, all partners should be provided with the summary tables shown in Fig. A2-3. The financial summary table is simply a total of cash paid into the partnership by each partner for the year, the cost of capital for the year (determined according to the definition in chapter 2), and the per hour costs of flying for each partner for the year (the sum of cash expense and cost of capital divided by hours flown). If you each pay fuel costs out of your own pocket, you each have to add this figure to your expenses run through the partnership account.

The summary table in Fig. A2-3. ties back into the financial model used to estimate the partnership's annual cost of flying. In fact, it is proof of the accuracy of your assumptions or the need for their revision. It shows how close you came to where you thought you should be financially at the end of the year. The engine and airframe reserve is equal to whatever reserve money was paid into the partnership account as part of the hourly payments made during the life of the partnership. Reserves put in during the year should equal the total hours flown during the year times the per hour reserve charge (you can calculate this amount and proof it to payments for hours flown actually received during the year). This amount is added to reserves already accumulated, and any assessments during the year are added to arrive at total reserves.

Note that since the partnership pays all of its expenses as they arise, and the only additional money put into the account is reserve money, the balance in the account at any time should be very close to the reserve amount (except for timing differences in checks issued by the partnership but not yet cashed).

A more informative financial statement

If you really are in the mood to track your actual costs of your flying compared to what you projected with this book's financial model, you can get a lot fancier in your financial record keeping (Fig. A2-4.). The income side differs from the more simple statement discussed above in that it lists hours flown and breaks out operating income separately. This is the per hour payment of each partner for hours flown and includes the reserves (but, again, not fuel if you each pay on your own). Thus you have up-to-date information on how many hours each partner has flown since the beginning of the year (cross-checked to the logs) and how much each partner has spent on operating and fixed expenses.

On the expense side, the keeper of the books assigns the appropriate expense category to each expense. At the end of the year each category is summed and each partner is assigned his or her share. The categories tie in with this book's financial model. Assumptions can be compared to results in great detail, and the costs of flying can be better understood as a result of experience.

A separate engine reserve account

Engine reserves may be kept as surplus cash in the regular partnership checking account, but it is a nice refinement of partnership accounting to keep a second, separate account strictly for the accumulation of the engine reserve. As payments are received for hourly flying charges, the partner keeping the books transfers the reserve amount of each hourly charge into the reserve account. A separate reserve account is a good idea because it protects the engine reserve from the temptation of alternate use. It can be a savings account earning a higher rate of interest than a checking account, and partners can tell at a glance how well their engine reserve is doing in comparison to the amount of hours the airplane is flying. They can more accurately evaluate whether they are reserving at a sufficiently high rate or if they are facing a potential special assessment at overhaul time.

Periodic financial reporting to the partners

It is important for the partners to be well informed about the financial affairs of the partnership. The partners need to know how much they have been putting into the pot and what expenses have been paid on behalf of the partnership. To this end the partners must be sent periodic financial statements, as well as proof that payments made by the partnership were for legitimate invoices. Monthly statements come to most partners' minds when the partnership is initially formed, but frankly, in most cases this is too time consuming. For the majority of partnerships quarterly statements are sufficient (perhaps sent just before quarterly meetings). For proof of payment of legitimate bills, copies of invoices over a certain amount (say $50) may be attached.

If the financial statements are diligently kept up to date, the checkbook is regularly balanced, and invoices are carefully filed, the quarterly mailing to the partners of the partnership's financial records should be a nonevent.

Safekeeping of Essential Partnership Documents

Partnerships will accumulate a fair number of essential documents. There is the partnership's founding document–the partnership agreement. There are a variety of documents related to the aircraft, such as the bill of sale, the title search results, and title insurance, and the annual insurance policy. There may be a loan agreement if the airplane is financed. While legally acceptable duplicates of these documents can be located, it is best to take good care of the originals in your possession. The best place for them is a fireproof safe. Make sure that all the partners have copies of all documents, but put the originals away in a secure place.

9

Pros, Cons, and Creative Options

The previous chapter completed the presentation of the partnership's elements. You should now be in a good position to set up a partnership of your own. How the guidelines presented so far can best serve your needs is largely a function of partnership size and the nature of the partners. There are pros and cons to partnerships of different sizes and special issues to watch for when setting them up. Let's look at some of the advantages and disadvantages inherent in partnerships of two partners, three or four partners, and larger partnerships.

Two Partners

If you can afford it, a partnership of two partners is best. Depending on the amount of flying you each plan to do, it can be almost like being a sole owner without the expenses of being one. It is also likely to instill a sense of mutual loyalty. You will each feel bad if you let the other partner down. There is no third or fourth partner to whom you can complain or rationalize. The two-person partnership will work very well right away, or it will not work at all.

There is a lot of room for being informal, for keeping structure and bureaucracy to the bare minimum. If you have similar flying objectives you can fly the heck out of the airplane as a team and get far more out of it than either partner would as sole owner. Or if you prefer, you can arrange the partnership in such a fashion that you hardly ever see each other and fly the heck out of the airplane individually, practically as sole owners.

The partnership agreement

The two-person partnership provides the greatest flexibility regarding the partnership agreement. The important objective is to set up an effective mechanism to resolve disputes and handle the dissolution of the partnership. Beyond that, it can be as minimal or as detailed as the partners prefer. "A schedule mutually agreed upon" is just fine if you know that the time slots you each want are complementary.

Finances can be handled in simple fashion. It is wise to include language that all expenses are due immediately and that no partner will pay anything on behalf of the partnership prior to receiving the other partner's share of the payments. But how strictly you enforce these terms is up to you. A periodic requirement to settle accounts is also worth including.

An often ticklish aspect of the two-person partnership is the issue of engine and maintenance reserves. Based on what they know of each other, two partners frequently trust each other to come up with the big bucks at maintenance and overhaul time, and surprise, surprise, someone can't deliver. Even the closest partners may want to set up a reserve account and fund it periodically based on the amount they fly. There is probably no need for nickel and dime checks to be sent in every time you fly around the patch, but you should both be obligated to make a deposit once the reserve amount you owe reaches a certain level.

A good plan for two partners is to sit down with the co-ownership agreement chapter (chapter 5) and go down the items one by one to decide the extent of detail required. A word of caution. If you don't know your partner very well, but necessity has brought you together and initial impressions are good enough to try a partnership, opt for greater detail.

In the end, the two-person partnership is truly a personal relationship, and no matter what a piece of paper says, ultimate success or failure depends on the nature of your relationship.

Operations considerations

Simplicity in operations is the desirable and easily attainable goal of the two-person partnership. Keep your fuel bills separate and keep good records of who flew when (through a logbook kept in the airplane), and usually all will be well.

The simplest scheduling method is the priority system, where each partner has the airplane exclusively for alternate weekends or particular days, whatever works best. Beyond that, informal notice can let the nonpriority partner know if the airplane is available. It is important for

Fig. 9-1. *A partnership to learn to fly.*

the nonpriority partner to be discreet. "I'll just hang out at the airport if you're not back by 2:37, but take all the time you want at the beach," might make the priority partner feel unduly pressured.

Maintenance considerations

The big maintenance questions are how much or how little to do yourself under the supervision of an A&P mechanic and what maintenance facility to use. You can learn a lot about the airplane working alongside an independent mechanic freelancing in addition to his regular job, and can keep the bills down, but there are also drawbacks. The mechanic may not always be available when you need him, so repairs and annual inspections may take longer than you would like.

Alternatively, a fancy maintenance shop that does everything from Jennies to jets will service your airplane thoroughly (though not always speedily, depending on backlog) for big bucks and no effort on your part except a few phone calls. The choice that places low demands on partner time but doesn't break the bank may be a shop that is not at a high-rent airport and has a reputation for specializing in light aircraft. Maintenance expenses are one of the biggest aircraft expense items, so even the most informal two-person partnership should have a clear understanding of committing to maintenance work only after mutual agreement.

Friction on maintenance issues can arise if one partner is an incorrigible tinkerer and the other one is terrified if anyone but an A&P mechanic lays hands on the airplane, even for preventive maintenance authorized under FAR 43.3(g) (see appendix 5). Again, either approach is fine; the real issue is partner compatibility.

Three or Four Partners

Three- or four-person partnerships are very popular because costs are greatly reduced yet the number of partners is sufficiently small to allow excellent aircraft availability. But when a partnership has more than two partners, a considerably more formal structure and organization is usually required. Demand for the airplane is higher, the chance for misunderstandings greater, and the administrative workload heavier.

The co-ownership agreement

The best advice for the three- or four-person partnership is to address everything in detail in the co-ownership agreement. Aircraft scheduling, expense billing and accounting, decision making, fueling, dispute resolution, the departure of partners, and partnership dissolution all need to be spelled out with great care. The relationship among the partners may be and often is very personal, but the partnership should be organized and run in an entirely businesslike fashion.

Operation considerations

While the three- or four-person partnership demands a certain degree of formal organization, it is still small enough to function effectively on a priority pilot basis if the partners' interest is to have access to the airplane for long stretches of time. A priority schedule can be laid out a month in advance and distributed among the partners. Some mechanism should be worked out to handle competition for the airplane among nonpriority pilots in case the priority pilot isn't using it. A first come, first served arrangement through positive contact with the priority pilot and a maximum nonpriority time limit usually does the trick.

If long-term use is not an objective, the partnership should have some simple scheduling arrangement, even if it is an informal phone check by a certain time of day with a partner designated to coordinate the schedule. For partners with Internet access, e-mail is an excellent alternative.

Fuel bills are still best kept separate, the partners being required to return the airplane refueled.

A responsibility that needs to be formally assigned to a partner is the partnership accounting. All partners should be kept apprised of their account activity by a periodic account statement (at least quarterly is recommended, monthly would be better). All partners should pay their share of any expense promptly, and a mechanism for specific partnership expense approval should be in place.

Separate bank accounts for engine and maintenance reserves are highly recommended. Periodic payments of accumulated reserves due is probably an effective means of collection, as long as the maximum limit on amounts outstanding is kept small.

It is an excellent idea to hold quarterly meetings with the express purpose of catching up on partnership matters, including the review of accounts.

Maintenance considerations

Maintenance considerations for the three- or four-person partnership are really no different from the two-person partnership. having a greater number of partners may provide more opportunity to do your own work on the airplane. The partnership may want to consider arranging periodic airplane washing and waxing outings.

Five or More Partners

Partnerships much beyond five or six members in size begin to assume the characteristics of a flying club. In fact, most insurance companies consider a partnership with more than six members a club. From a legal standpoint, it is advisable for larger partnerships to incorporate. The airplane is the corporation's single asset and the members can retain equal percentage shares in it as in the traditional partnership.

Organizationally a larger partnership needs to assume some of the more structured characteristics of a flying club. If it is incorporated, at least on paper, it will have to have a minimal number of directors (though the group will still be small enough to make decisions through universal participation). More importantly, the workload of running the partnership will require the formal assignment of more duties to specific partners. A formal scheduling system needs to be set up and run, accounting becomes a more labor-intensive responsibility, and the group may need to formalize checkout and currency requirements. The partnership agreement has to cover all aspects of the organization in great detail, and may need to be supplanted by operating rules.

Larger partnerships are feasible and are especially popular as a means of providing an opportunity to fly more sophisticated equipment, such as a Bonanza or a light twin. But such partnerships require the adoption of at least some of the characteristics of a flying club (see chapter 11).

Be Creative

So far I have focused on how to go about methodically assessing all aspects of the partnership alternative and how to get the most equipment for your money. The focus has been on the typical production singles and light twins of the average pleasure flyer's world. But there are other, even more creative applications of the aircraft partnership concept.

How about forming a partnership to learn to fly, to build and fly a homebuilt aircraft, or to fly a warbird. These are all feasible, offer financial advantages over the traditional routes for engaging in these activities, and are presented as examples of what is possible with a little imagination. The examples are meant to stimulate your mind to come up with your own creative alternatives and solutions. Take the numbers not at face value, but as an approximation. Collect the data applicable to your own environment and do your own in-depth analysis.

A Partnership to Learn to Fly

Well, why not? If there are several of you who have put off learning to fly because of earthly constraints, but are serious about correcting this grave error, you should certainly take a good look at owning a trainer in partnership as an alternative to learning on rental airplanes. The basic idea is this: you and your partners buy an older, but mechanically sound trainer for as little money as possible, learn to fly on it, and as soon as you all have your licenses you sell it for about as much as you bought it for. Will it cost less than the rental alternative? Only the numbers will tell.

A prerequisite for making such a partnership work is the commitment of all partners to be serious about learning to fly. You must also be realistic about your assumptions regarding the time it will take you to get your pilot's certificate, and you must also buy the airplane. Its price must not be above the current market price or you will have a hard time recouping your investment when you sell it. To further protect your investment, you are best off choosing a model that has a high resale value history.

Another essential requirement is the reliable mechanical condition of the airplane. A hangar queen is never fun to own, but can be especially detrimental to your progress when you are learning to fly and continuity is of the essence. The airplane should have plenty of hours left on the engine to cover your needs and still be salable at the end of your lessons. Buying one with a fresh annual inspection, or having an annual inspection performed at the outset is highly recommended. You don't want the airplane in the shop just when you are about to solo.

Don't be concerned with cosmetics. Many perfectly sound trainers have worn paint jobs. Pretty paint will only add to your costs and can wear off while you own the airplane, decreasing the value of your investment. Fresh, pretty paint on an airplane for sale can also be an attempt to gloss over darker secrets awaiting the bright-eyed, first-time buyer. Which brings us to another important point. How to go about buying the airplane.

It is traumatic enough to be a first-time airplane buyer when you already have a license and some experience with airplanes, but it can be a real trap if you have no experience at all. To minimize the chance for surprises, there is only one alternative. Seek the assistance of someone experienced in buying airplanes whom you can fully trust (see chapter 12).

A good plan is to make a package deal with an instructor to give you your lessons. Shop around among the many instructors who contract directly with airplane owners, and try for a discount (including a good deal on ground school) by committing the partnership to the instructor for all training needs. If you make such an arrangement, the instructor may also be able to help in selecting the airplane.

Let's take a look at some numbers. Cessna 150s are the stereotype trainer and retain their value well. They have also been around for decades, so the selection is wide. Let's say that three of you want to buy one to learn to fly. First you will have to make a realistic assumption of how long it will take each of you to earn your licenses. Then you will have to see how much it would cost under the rental option and compare this figure to the cost of the partnership alternative.

The regulations allow you to get a private license in as little as 35 hours, but this figure is generally believed to be unrealistic if you can't fly practically every day. Scheduling constraints and weather delays will increase the number of hours most students will need. An average time often mentioned is 75 hours. The availability of your own airplane should give you the flexibility to keep at it consistently, so let's use 75 hours as the time each member of your partnership would require. At $80 per hour for a dual at a rental establishment, the private pilot's license would cost each of you $6000. This is the figure to beat.

Let's use the financial analysis model to see how much 75 hours per partner would cost in a three-way Cessna 150 partnership (see Table 9.1). The purchase price of the 150 is $15,000, requiring $5000

	Total Cost	3 Pilots, per pilot
CAPITAL INVESTMENT	15000.00	5000.00
ANNUAL FIXED EXPENSES		
Tiedown/Hangar	960.00	320.00
Insurance	750.00	250.00
State Fees	120.00	40.00
Annual	450.00	150.00
Maintenance	900.00	300.00
Cost of Capital (non-cash)	1,050.00	350.00
Total Fixed Expenses / yr	4,230.00	1,410.00
HOURLY OPERATING EXPENSES		
Fuel	13.00	13.00
Oil	0.31	0.31
Engine Reserve	5.00	5.00
General Maint Res	2.00	2.00
Total Op Exp / hr	20.31	20.31
TOTAL HOURLY EXPENSES		
75 Hours		39.16
TOTAL ANNUAL EXPENSES		
75 Hours, Own		2,937.25
Plus instructor @ $20.00*75		1,500.00
Total License Cost		**4,437.25**
Total License Cost, Rental		**6,000.00**
Savings per pilot		**1,562.75**

Table 9.1. *A trainer partnership to learn to fly.*

per partner. From beginning to end (including finding, buying, and selling the airplane) it would comfortably take you at the most a year, and probably less to earn your licenses, so it is assumed that you hold the airplane for 1 year. To keep the example simple I've used a rate of $20 per hour for the instructor for the entire 75 hours. I've ignored ground school costs because they are incurred in both alternatives, so they are a wash.

Look at the last three lines of Table 9.1. The savings are considerable, assuming that you can sell the airplane in 1 year with another 300 hours on it for the same price you paid for it. If you bought well, this may be possible. But even if you take a $2000 haircut on the total resale price, your net savings over the rental alternative are reduced only by $666 per person, which still puts you ahead by $895 ($1562 – $666).

In this example the partnership didn't borrow to buy the airplane. If you finance 50% ($7500) of the airplane, each partner would have to come up with an initial investment of only $2500. Annual payments on a loan of $7500 for 10 years at 10% would be $396. Without a resale haircut, the savings would be $1166 per partner ($1562 – $396). With a $666 loss per partner on the resale of the airplane, each partner would still be $500 ahead on a net basis ($1166 – $666).

The savings could be larger if the partners completed their licenses and the sale of the aircraft in a shorter period of time. Tie-down fees would be less, and up to a point, there would be an insurance refund if the whole year's worth of insurance is not used.

There is a very wide range of prices and equipment alternatives out there, so do your homework and be careful. But with a little creativity and perseverance you may find the alternative of buying a trainer in partnership preferable to renting one on your own.

A Partnership to Construct a Homebuilt Aircraft

It is a fascinating thought to fly in something built by your own hand, and in the minds of most pilots that is exactly what it remains—a thought. Building an airplane alone is an immense job requiring a lot of technical skill, access to a suitable workspace, and a good set of tools. Among those who try, the completion rate is not that high. Then there is the problem of money. Some homebuilt kits, excluding the engine, can cost as much as a good, used, production four seater.

If you can't go it alone but are determined to construct an airplane, consider a partnership to build one. In many respects, such a venture is no different from a regular airplane partnership.

Compatibility with your partners is just as important, perhaps even more so, considering that you are not only sharing a piece of equipment, but actually have to work together and rely on each other to complete a variety of fairly demanding tasks. Once the airplane is complete, the partnership becomes just like any other. What has to be sorted out at the outset is how it will function during the construction process. The big risk is that you are making a considerable financial and nonfinancial commitment to something that doesn't even exist. You have to have a strong belief that the project will be completed. You should be able to see a path that will get you there, and you should also be prepared to handle setbacks. When you build an airplane on your own you have only yourself to blame for not living up to your goals. In a partnership there are always the partners, and the finger pointing can easily escalate out of control when the wheels miss the wheel wells by a fraction of an inch.

A good way to help you and your partners focus on the task, the responsibilities, and the pitfalls is to work out a partnership agreement for the construction phase of the venture just as you would for a regular aircraft partnership. In broad scope your aircraft construction partnership agreement should cover at least the following points:

- *Detailed specifications of the aircraft.* Spell out the design and model to be built, the plans to be used, the kit or prefab components to be used, the engine (specifications, used or new, how many hours), and the accessories and instrumentation to be installed.

- *Costs.* Have as detailed a breakdown of the costs as you can. In the case of a kit-built airplane, this task is fairly easy. For scratch-built aircraft the costing of the raw materials may be more difficult if you are not sure how much stock you will end up using and how much wastage you will have. A problem with costs is estimating future costs of components you will not need until much later in the building process. Considering that some projects can take many years to complete, there is a lot of room for error.

- *Construction schedule.* Map out, as any businessperson would, the completion schedule of your project. Set target dates for phases to be completed. This requires a good understanding of the labor and time requirements of your project.

- *Funding.* Establish each partner's financial commitments and when such commitments are due. Usually the big expenses are paid stage by stage as each major phase of construction is begun. The partnership should also have an expense tracking system in place and limit the amount that each partner can expend on behalf of the partnership without the others' approval.

- *Cost overruns.* Higher than anticipated costs can be a big bone of contention, and the partners should have an understanding of how they will be handled.

- *Commitment of partner time.* Questions about who spent how much time on the project can sometimes cause friction. It is worthwhile to agree beforehand to some amount of time commitment per partner. Some partnerships make this simple by making each partner wholly responsible for completing certain parts of the project, whatever time it takes.

- *Sweat equity.* One common reason for forming airplane construction partnerships is because one partner has the money and the other partner knows how to build airplanes. If any partner is going to contribute labor in exchange for a share of the airplane, the terms and measure of that labor must be carefully defined and well understood by all the partners.

- *Conflict, dissolution, and arbitration.* The ticklish question of how to deal with disagreements and how to dissolve the partnership if a partner wants out for whatever reason must be addressed. A half completed miniplane doesn't do well at yard sales, which is why you have to have a good idea of your ability to see the project through before you begin.

These are just some of the issues you should be thinking about. Undoubtedly there are many more peculiar to your own circumstances. Be sure to run any agreement by your lawyer before you sign on the dotted line.

The idea of forming a partnership to build an airplane is a good one. The rewards can be immense, both in the sense of personal satisfaction and also in aircraft performance. And when at three o'clock in the morning you are looking for that rivet gun you just know your partner was the last one to use, think of what that first landing at Oshkosh will feel like.

A Warbird Partnership

How does the sound of 32,000 hp strike you? To some it's just a lot of noise, but if you are a warbird fan it may be music to your ears. Especially if emitted by growling, snarling radials as was the case during a flyby of over 50 T-6 Texans at Oshkosh some years ago. To many pilots watching from below the old warhorses seemed as

Fig. 9-2. *A Warbird partnership for Chuck Yeager wannabes—the spunky L-29 Delfin, ex-Warwaw Pact Trainer.*

inaccessible as an F-16, but that might be a mistaken impression. It doesn't take nearly as much as you may think to be up there flying one as an owner. Although warbirds will never be inexpensive, a partnership will dramatically cut the cost of flying one of these exciting planes and put them in reach of someone accustomed to being a sole owner of a Saratoga or a Bonanza.

Warbird prices can run into the millions for rare, well-restored World War II bombers such as the B-17, and a P-51 Mustang can go for as much as $500,000. But the lighter iron can be surprisingly reasonable by warbird standards. Some vintage jets, such as the Fouga Magister, Willy Messerschmitt's Saeta, and the L-29 Delfin go for around $100,000 to $150,000 in good condition, and a T-6 Texan can be had for under $100,000. So why not trade in that Bonanza for a Texan?

The big bugaboo is operating cost, as well as the matter of getting good training. Even the less expensive warbirds consume enormous amounts of fuel, maintenance is hideously expensive, and parts may be hard to come by. But if you get a good price, and are willing to spread the costs around through a partnership, a warbird may be not out of reach.

Consider the T-6 Texan. The numbers presented in Table 9.2 are only examples (actual costs may vary widely depending on circumstances and changing prices), but give a good ballpark idea of what a prospective T-6 pilot might face.

The T-6 in the example is in good average condition and goes for $95,000. Let's be daring and compare the T-6 partnership to the expense of owning a Piper Arrow II alone. Key figures are the total hourly expenses for the number of total hours flown (100, 150) and the total annual expenses. Let's look at 100 hours, since you would rarely want to own an Arrow alone if you were going to fly it less than that, and for safety reasons you shouldn't be flying a Texan if you can't fly it 100 hours a year. The Texan costs $291 per hour if owned alone; the Arrow costs $111. That is a big difference. Now look at four partners in the Texan. The rate drops to $161 per hour. That is only $50 per hour more than the Arrow, or, on a total annual basis at 100 hours, $5000 more.

You may think that an extra $5000 is a lot to come up with, but consider this: You have $45,000 in the Arrow. A one-quarter share

			T-6				Arrow
NUMBER OF PILOTS	1	2	3	4	5	6	1
CAPITAL INVESTMENT	95000.00	47500.00	31666.67	23750.00	19000.00	15833.33	45000.00
ANNUAL FIXED EXPENSES							
Tiedown/Hangar	2,700.00	1,350.00	900.00	675.00	540.00	450.00	960.00
Insurance	4,000.00	2,000.00	1,333.33	1,000.00	800.00	666.67	1,500.00
State Fees	120.00	60.00	40.00	30.00	24.00	20.00	120.00
Annual	2,000.00	1,000.00	666.67	500.00	400.00	333.33	750.00
Maintenance	2,000.00	1,000.00	666.67	500.00	400.00	333.33	1,500.00
Cost of Capital (non-cash)	6,650.00	3,325.00	2,216.67	1,662.50	1,330.00	1,108.33	3,150.00
Total Fixed Expenses / yr	17,470.00	8,735.00	5,823.33	4,367.50	3,494.00	2,911.67	7,980.00
HOURLY OPERATING EXPENSES							
Fuel	80.00	80.00	80.00	80.00	80.00	80.00	21.00
Oil	2.00	2.00	2.00	2.00	2.00	2.00	0.13
Engine Reserve	25.00	25.00	25.00	25.00	25.00	25.00	7.50
General Maint Res	10.00	10.00	10.00	10.00	10.00	10.00	3.00
Total Op Exp / hr	117.00	117.00	117.00	117.00	117.00	117.00	31.63
TOTAL HOURLY EXPENSES							
100 Hours	291.70	204.35	175.23	160.68	151.94	146.12	111.43
150 Hours	233.47	175.23	155.82	146.12	140.29	136.41	84.83
TOTAL ANNUAL EXPENSES							
100 Hours, Own	29170.00	20435.00	17523.33	16067.50	15194.00	14611.67	11142.50
Own (cash only)	22520.00	17110.00	15306.67	14405.00	13864.00	13503.33	7992.50
150 Hours, Own	35020.00	26285.00	23373.33	21917.50	21044.00	20461.67	12723.75
Own (cash only)	28370.00	22960.00	21156.67	20255.00	19714.00	19353.33	9573.75

ASSUMPTIONS:	T-6	Arrow II		T-6	Arrow II
Aircraft Value:	95000.00	45000.00	Maintenance/yr:	2000.00	1500.00
Hangar/Month:	225.00	80.00	Gen Maint Res/hr:	10.00	3.00
Insurance/yr:	4000.00	1500.00	Engine Reserve/hr:	25.00	7.50
Annual:	2000.00	750.00	Time Before OH (hrs):	1200	2000
Fuel Cons (gal/hr):	40.0	10.5	Engine MOH Cost:	30000.00	15000.00
Fuel Cost ($/gal):	2.00	2.00	Cost of Capital Rate:	7.00%	7.00%
Oil Cons (qt/hr):	2.0	0.13	State Fees/yr:	120.00	120.00
Oil Cost ($/qt):	1.00	1.00			

Table 9.2. *Financial analysis of a warbird partnership.*

in the Texan is only $23,750. That is a whopping $21,250 less than what you already have in the Arrow. Another way to look at it is that the excess investment in the Arrow is the source of the extra $5000 per year for 4 years. Sure, by the end of the fourth year it would be gone, but you would have 400 hours of T-6 time and countless adventures in your logbook. And at the end you will still have your one-quarter share in the Texan. Since they don't make Texans any more, it will most likely have held its value. You can always sell it, and having gotten used to partnerships, buy a half share of another Arrow and fly it 100 hours a year, just as you were doing before your wild and woolly Texan days. If there is a will, there is always a way.

The importance of training

Before you do sell that Arrow and jump into a Texan, take one point to heart. It is absolutely imperative that you get the appropriate professional training to fly a warbird before you blast off in one alone. That goes not only for soloing it, but also for all other types of operations such as maneuvering, dogfighting, and aerobatics. These machines are a joy to fly, but they were built for war, to be flown by highly trained military pilots. Your standards should not be anything less.

Case Study: Ximango Motorglider—Two Partners

The aircraft

The two-seat Ximango motorglider is a very capable hybrid aircraft. Powered by its 81-hp, four-cylinder Rotax aircraft engine it exceeds the performance of a Cessna 152. In glider mode, with the engine shut off, it has a glide ratio of 31:1, which means that in still air it can glide 31 miles from an altitude of 1 mile. As a glider its performance equals that of medium performance recreational/training gliders, giving it respectable cross-country performance.

The Ximango's great advantage over pure gliders is that it needs no tow plane to launch it and when the pilot runs out of lift in glider mode the engine can be turned back on, converting the aircraft in midair into a power plane. If conditions are not conducive to soaring or the pilot simply needs a power plane, the Ximango can be used just like a Cessna 152 or a Katana. Based on a French Fournier design, the Ximango is made in Porto Alegre,

Case 9-1. *A Ximango motorglider is most efficiently put to use in a partnership.*

Brazil. The Ximango attracts pilots from both the glider and power plane community who want to put the fun back into flying. The Ximango of this partnership was purchased new. The partners have had the aircraft for 3 years.

Partnership background

This is a two-person partnership. Both partners own other airplanes (one an A-36 Bonanza) and have access to gliders. The partners knew each other for a long time as glider pilots before forming the partnership.

They bought the Ximango new, strictly as a fun airplane, so that they could indulge in recreational soaring and power flying at a moment's notice.

Partnership structure

The two partners chose to incorporate as a limited liability corporation. The main reason for incorporation was to benefit from the liability protection this form of ownership offers. The by-laws of the corporation serve as the partnership agreement, though there is very little detail in it beyond the minimum language required to make it a legal document. Elaborate rules and provisions were avoided because the partners know each other well and fly the airplane relatively little on mostly short, local flights.

Finances

Each partner bought a 50% share of the Ximango. Neither borrowed any money to buy the airplane.

Insurance and the hangar costs are the main fixed expenses of keeping this airplane. Considering that it is mostly operated in glider mode, there are very few hourly operating expenses except for the cost of gas. Gas consumption at 4 gallons per hour is minimal. Fuel expense per flight hour is especially low since the engine is shut down for a good portion of most flights. Because of these characteristics, and the fact that they each fly the airplane about the same amount of hours per year, all expenses except gas are shared equally by the partners.

Operations

The partners maintain a joint checking account for the aircraft. They each put in an equal lump sum to fund the account and periodically top it off with equal contributions when it gets low. One partner pays all the bills and copies the other partner on the checking account statements, which serve as the financial records of the partnership. Gas is the only unshared expense, which each partner buys individually as needed.

The partners do not formally set aside an hourly engine reserve. Should an overhaul or any other major expense become necessary, an assessment would be made to pay for it.

Scheduling is informal because of the nature of the flying the partners do in the airplane. If the Ximango is taken for more than a local flight the partners check with each other beforehand, but given that it is a secondary airplane for each partner, it is rarely taken for any length of time.

Maintenance

There is minimal maintenance to be done on this airplane, but what little there is gets done by the dealer, who is located near the partners

The partnership experience

The partnership has worked well. There have not been any points of contention. The partners attribute the smooth experience to their mutually accommodating nature and the simple, straightforward structure of the partnership. Each partner flies approximately 50 hours or less per year and agrees that neither would see any financial sense in keeping the Ximango on their own.

10

Joining an Established Partnership

We've all seen the ads: "1/3rd share of Skylane, $18,000, based BVY, $39 wet, good availability, call. . . ." It is a tempting alternative to join an already existing partnership instead of going through the trouble of finding potential partners and setting one up from scratch, and if the opportunity pans out it may be the perfect solution. But the joining partner needs to pay careful attention because in some respects joining a partnership already in existence can be even trickier than forming one.

When a whole new partnership is formed everyone sets out on an equal footing. The partners may know each other quite well, but the habits and ways of the partnership are yet to be established. There is plenty of room for give and take all around. However, when you join an existing partnership, you are in a sense intruding on a relationship that has been going on for some time. Chances are it has developed little nuances and has become comfortably set in patterns to which you will have to adapt or which will have to be adjusted. If forming a partnership is like getting married, then joining one already in existence is like joining your in-laws' in their family business.

No matter how good the intentions are of the partners taking you in, the fact remains that you will be the new kid on the block. The partners will be paying close attention to you, and on a subconscious level at least, they will be waiting for you to prove yourself, to give signs that you really deserve to be one of the group. This will certainly be the case while you are negotiating to join the partnership, and will most likely linger for some time after you join until the partnership's patterns settle once again into a familiar groove.

It can be quite intimidating to be under a partnership's magnifying glass. Such scrutiny may cloud the joining partner's thought process in evaluating the partnership and may even pressure the joining partner into conclusions or decisions with which he may not be entirely comfortable. This need not be so. A little assertiveness and self-assurance backed by a good dose of quiet competence will make joining a partnership as smooth as forming one.

Once you have defined your personal objectives and financial commitment your task will be twofold. You will have to closely scrutinize the partnership's aircraft and you will have to carefully evaluate your potential partners. To some extent mutual interests will be at work, because your partners will also want to make the right decision. But misperceptions can arise, even with the best of intentions. Let's take a closer look at how to go about joining a partnership.

Checking Out the Aircraft

The evaluation of the partnership's aircraft should be the joining partner's most straightforward task. It should be a pure business decision, deliberated in as much detail as if the aircraft were being purchased to form a new partnership. Yet it is in the evaluation of the aircraft that a joining partner is most easily badgered by the members of the partnership.

For most people their airplane is a matter of deep personal pride. It is only natural for the members of a partnership to view a critical outsider with suspicion and to be defensive. On the other hand, a potential joining partner is reluctant to behave as if he were bargaining at the horse thieves' market.

At the market and in a normal sale of an airplane, once the mutual bullying is over and the bargain is struck, buyer and seller say good-bye and may never see each other again. But should the joining partner decide to buy, he will be moving in with the sellers. The whole lot of them! So, just to get a better deal, he will be wary of loudly declaring that his potential partners' alleged cream puff is a bucket of bolts that would make the Wright brothers get back into bicycles. This wariness is misplaced. The joining partner should place concerns about hurting the owners' feelings firmly behind the task of objectively evaluating whether or not the airplane is a good buy.

As the joining partner, your first order of business is to get all the pertinent details on the airplane and the financial terms of the sale. Evaluate this information in light of the research you have done, according to the guidelines offered in chapter 12. Is the price in the

Fig. 10-1. *You never know when a Mooney 252 partnership may have a vacancy.*

ballpark? Does the airplane have all the features, characteristics, and equipment you set as your objective? Is it in the age group you were looking for? Can you live with the number of hours on it? These are all routine questions; make a list.

If the preliminary information checks out, proceed with inspecting the airplane. Follow the guidelines set out in chapter 12. When you are joining a partnership it is especially tempting to rely on the partners' mechanic, or even soothing statements from the partners themselves. After all, won't they continue to fly the airplane right along with you when you join the partnership? Sure they will, but if the engine is about to blow up and will require an overhaul in the next few months, their share of the bill will be that much less if you join. Consider your share of a $15,000 overhaul in a three-way partnership before you lamely accept your future partners' assurances.

The only way you will know for sure is by having the aircraft checked out–by your own mechanic.

There is another reason to get your own mechanic. Inevitably, psychology is at work during the transaction. You may end up feeling pretty lonely and pressured when you show up by yourself to face two or three partners and their mechanic. Appearing with your own mechanic in tow puts you on a more equal footing. It shows that you really mean business and know what you are doing.

It should be easy to arrange for the assistance of a mechanic you know or who has been recommended by people whose opinions you trust. Conduct the prepurchase inspection just as you would if you were buying an airplane to form a brand new partnership. Follow the guidelines in chapter 12.

Fly the airplane. Better still, have one or more of the partners fly it and observe their skills and habits. When the inspection is complete and you have come up with a list of squawks, you have three options. You can get the partners to reduce the price by your share of fixing the problems. You can specify in writing that the squawks listed are to be fixed at the expense of the original partners. Or you can walk away.

A worthwhile option, if the timing is right, is to participate in or observe an annual inspection done on the partnership's airplane just before you buy into it. Besides teaching you a lot about the airplane's mechanical condition, the annual inspection will also tell you a lot about the partnership. Note how the partners handle problems, how much of their own work they do, how much maintenance they defer, and most importantly, how their approach to maintenance fits in with yours.

If the airplane checks out and the price is right, it is time for the joining partner's most challenging task, evaluating the partnership.

Evaluating the Partnership

As in most situations where human emotions are a factor, the evaluation of a partnership is a subjective process. In the end you click or you don't. It is fine for the members of the partnership you are about to join to be laid-back good ol' boys if you are too. Or they can all be computer nerds if you are too. The trick is to find the common link.

In spite of human emotions and personalities, there are objective factors you can evaluate which will help you decide whether or not to join the partnership. A good place to begin is the partnership agreement.

Fig. 10-2. *The Warrior is the low-wing equal of the Cessna 172.*

Read the partnership agreement carefully and see how well it fits with your ideals. A good partnership agreement is a promising sign. Take note of any sections particularly important to you. Scheduling, the sharing of fixed expenses, fuel purchases, decisions, and the sale of partnership shares come to mind. If something is not to your liking, see how willing the partners are to make amendments.

Next, evaluate how the partnership practices what it has set for itself in the partnership agreement. The greatest agreement is worthless if not put into practice. During your checkout of the airplane you will have a chance to observe the partners in action, especially if you fly with one or more of them, or participate with them in an annual inspection.

Try to get a handle on how orderly or casual they are around the airplane. What preflight habits do they have and expect from you? Do they insist on checklists or do they rely on memory? If the latter is the case, how good is their memory? Do they consistently practice particular operating procedures in flight? Do they show a tendency to believe in one "right" way of doing things, or are they keen to explore and allow the practice of legitimate alternatives? What post-flight procedures do they follow? How do they expect to find the aircraft when they come out to fly?

Discuss with them at length their use of the aircraft. Do they really take turns flying it away for days on end, or do they all fly to the coffee shop 50 miles down the 233 radial every weekend?

Examine carefully how they keep track of flight time and expenses. Review all the flight records. The partnership flight log can give you insight into the group's flying activity. Ask to see the partnership financial records and review them carefully. Does the checkbook balance? Are the records accurate? Are all the partners regularly informed of the partnership's financial transactions? Do all partners appear to pay their share of expenses in a timely fashion? Does the partnership have a separate engine reserve account? Is enough being reserved per hour?

As long as regulations and safety margins are observed, there is really no right or wrong way to run a partnership, and that is not the question to ask when evaluating one. What ultimately makes any partnership work is that all of its members are comfortable with the style in which it is organized and run. As a joining partner, your big concern besides the airplane should be how well you fit in. Beyond checking out the facts, the best way to get to know your potential partners is to hang out with them for a while before you make a final decision. And remember, they will be as keen to make the right move as you are.

11

When Is a Partnership a Flying Club?

In concept the flying club doesn't differ all that much from the airplane partnership. Like the partnership, it is a gathering of likeminded pilots who want to bring the cost of flying down to a level they can afford by engaging in a cooperative flying arrangement. A good way to think of a flying club is as a very large partnership. The chief financial advantage of a flying club is that its members have to come up with far less capital to be in business than the members of a small partnership.

Depending on membership size, the flying club may also be in a better position than most partnerships to provide a fleet of different aircraft for the use of its members. It is not uncommon for even relatively small flying clubs to have two aircraft, one a modest VFR runabout for local recreational hops, the other a high-performance complex airplane for serious traveling. This is an excellent arrangement for more efficiently matching aircraft capability with flying needs, but it is financially out of reach for most partnerships.

A worry for pilots considering joining a flying club as an alternative to a partnership is aircraft availability given the large number of club members. A generally accepted rule of thumb is that a club should have no more than 15 members per airplane to ensure an acceptable level of access to all members.

The greater number of pilots involved in a flying club also makes it difficult to operate along the relatively simple and informal lines of a partnership. A higher level of organization and operating rules are required to make the flying club work.

As the number of pilots increases in a cooperative flying venture, it becomes impractical to make all decisions by consulting all the members and arriving at a consensus or a majority vote. Some formal delegation of responsibilities is necessary to efficiently conduct the group's day-to-day business and make decisions on the group's behalf. The group needs to elect a board of directors to oversee it and make decisions on its behalf and appoint officials from within its ranks to perform such functions as president, treasurer, operations officer, maintenance officer, and chief pilot.

A small flying club of 10 to 15 members can elect the board of directors to also serve as the officers of the club. In larger flying clubs it is more practical for the members to elect a board of directors who then have the authority during their term to appoint the officers as and when needed.

In order for the officials to effectively perform their duties, to allocate flying opportunities equitably, and for all members to clearly understand the operating procedures and their rights and responsibilities, the flying club needs more comprehensive organizational and operating rules than a partnership. These rules are set forth in the flying club's by-laws and its operating manual, which should be issued to all members.

If this description has a corporate ring to it, that is because the flying club is essentially a small corporation, and as such it is best incorporated. As is the case with partnerships, maximizing liability protection is another powerful incentive to incorporate the flying club.

If you are organizing a large cooperative flying organization and are not quite sure whether to establish it as a partnership or flying club, your insurance company will make the decision for you. Most insurance companies consider any cooperative flying organization with more than six members a flying club and will expect to see it organized as one, complete with by-laws. This should not cause a problem if your group is small enough (6 to 10 members) to continue to operate largely in the style of a partnership. A good partnership agreement is easily converted into by-laws with a few minor definitional adjustments. As your group grows and additional organizational structure is required, you can adjust the by-laws to reflect the changing conditions.

The remainder of this chapter examines the differences between a partnership and a flying club in terms of financial structure, financing and insurance, legal and organizational structure, and operations, safety, and maintenance.

Financial Structure

The components of the flying club's financial structure are not all that different from those of the airplane partnership. As in a partnership there is the cost of buying the aircraft and the fixed expenses and operating expenses required to operate the aircraft. The challenge for the flying club is to structure the payments it receives from its members to predictably meet these expenses. Here's how its done.

Buying the aircraft

Just like the pilots in a partnership, the members of the flying club buy their own share of the aircraft, but they do this by buying a share in the club, which owns the aircraft. If a group of 15 pilots wants to start a flying club by purchasing a Cessna 172 that costs $35,000, each member would have to come up with $2333 to buy a share in the airplane, if it were paid for in cash. When a member leaves the club, his or her share is salable according to the terms of the club's by- laws.

If 50% of the purchase price were borrowed, each member would have to come up with only $1167, and would have to pay his or her share of the loan payments (a component of fixed expenses). It would be very easy to get a bank to provide a loan for 50% of the purchase price of the aircraft for, say, 10 years at 12%. Under those terms, the annual loan payment is $3013, which amounts to only $201 per year per club member if there are 15 members. On a monthly basis this is only $17 per member that would have to be added to fixed expenses.

One could say that even $1167 is a lot of capital to come up with for someone who is looking for the least expensive flying option, and the more mature flying clubs have lower entry fees. I'll address later how they accomplish this.

Fixed expenses–monthly dues

The fixed expenses per aircraft (tie-down, insurance, etc.) are predictable, and payable, as you may recall, whether or not the airplane flies. The total fixed expense per year is the bill the flying club knows it has to pay no matter what, so the club has to have a predictable (a "fixed") source of income to cover this expense. The solution is to require that each member pay a fixed sum to the club in the form of monthly dues. If the total annual fixed expenses for the Cessna 172 are $6500 per year, each member has to come up with $433 in annual dues, which is only $36 per month.

Fig. 11-1. *This P-210 and Saratoga are on the flight line of a unique flying club.*

The flying club will also have some additional fixed expenses over and above the expenses directly linked to its aircraft, especially if it is a large club. Typical expenses in this category are office rental, stationery, accountant's fees, and the cost of administrative help. These general fixed expenses also have to be taken into account and covered by the annual dues. Increasing the dues of the 15-member Cessna 172 club in our example from $36 to $45 per month yields an additional annual income to the club of $1620.

The flying club has another fixed expense that the typical partnership doesn't have–depreciation. Depreciation is a noncash expense which the flying club can take because it is run as a business. Simply put, an asset such as an airplane is assumed to lose value each year. A business can deduct this loss of value from the income it receives for the year and is not taxed on this amount; it remains in the company as a reserve available to replace the asset when it is fully depreciated. The rate and the number of years over which an asset can be depreciated is set by the Internal Revenue Service and should be familiar to the flying club's accountant. It may or may not be advantageous to allocate depreciation expense into the annual dues figure, depending on how much net income the club has to report to the IRS; check with your accountant.

The noncash cost of capital, which is an important fixed cost to each partner in a partnership, may be ignored in the case of the flying club because the per member capital contribution is so low that its cost becomes negligible.

Operating expenses–hourly rental rate

The costs of operating the aircraft per flying hour, as you may recall, are incurred only when the airplane flies (fuel, oil, engine and maintenance reserves, etc.). As is the case in the partnership, these costs are paid in full by the member flying the aircraft in the form of an *hourly rental rate*. It is especially important for the flying club to accurately estimate engine and maintenance reserves because club members tend not to look at the club as co-owners but as users of its services (clients) and are less inclined to look favorably on special assessments than a group of partners. The good news is that even with generous engine and maintenance reserve rates per hour, the flying club still tends to be the best deal in town.

Surplus–the key to growth

As a flying club grows and matures it can slowly but steadily build a healthy surplus reserve by building a profit into its dues and hourly rental rates. Spread over a large number of members, such a cushion is financially painless per member, yet when it is added up it amounts to a substantial sum. An extra $10 per month in dues per member in a club with 45 members amounts to $5400 extra per year. If each of those members flies 25 hours a year, an extra $10 per hour in aircraft rentals amounts to $11,250 per year. The two surplus sums together amount to an additional income of $16,650 per year.

This surplus can be used to pay down the loans used to buy the airplane, as well as to repay any initial investment made by the founders of the club. This will enable the club to eventually own aircraft free and clear, acquire additional aircraft, and reduce the entry fees charged to new members, making flying even more affordable. However, for the numbers to work, the flying club's biggest challenge is to keep the membership at a sufficiently high level to assure it of the necessary income. Take a look at Table 11.1 for a comparison of the flying club's financial structure to that of a partnership. Table 11.1 compares a three-person partnership to a club with two airplanes and 30 members.

The flying club advantage

Compared to a partnership, the flying club will always be the less expensive option because the expenses are spread over more pilots.

FLYING CLUB BREAKEVEN FINANCIAL ANALYSIS AND COMPARISON TO PARTNERSHIP AND RENTAL OPTIONS

EXPENSES ($)	TOTAL	AIRCRAFT #1 PA28-201 500 <Hr/Yr	AIRCRAFT #2 C-172 500	Piper Archer Three Pilot Partnership $/pilot	
ANNUAL FIXED EXPENSES					
AIRCRAFT RELATED					
Tiedown/Hangar	2,040	1,020	1,020	340	hangar
Insurance, Aircraft	4,000	2,500	1,500	500	insurance
Fees and Taxes	240	120	120	40	state fees
100 Hour and Annual Inspections	3,500	2,000	1,500	250	annual
Maintenance (paid through reserves)	0	0	0	334	maintenance
Loan Payments	0	0	0	0	loan pmt
Depreciation	4,000	2,500	1,500	1,167	cost of capital
GENERAL EXPENSES					
Office rent	0				
Accountant's Fee	250				
Manager's Salary	0				
Mechanic's Salary	0				
Supplies/Administrative	250				
Other	0				
Total Fixed Expenses/yr	14,280	8,140	5,640	2,631	total fixed exp/yr
HOURLY OPERATING EXPENSES					
Fuel	50	20	14	16	fuel
Oil	3	1	1	1	oil
Engine Reserve	27	12	5	10	eng. res.
General Maint Res	16	8	3	5	maint res
Total Op Exp/hr	96	41	23	32	tot op. exp/hr
Total Op Exp/yr	32,000	20,500	11,500	58	$/hr @ 100 hr/yr
TOTAL ANNUAL EXPENSES	46,280	28,722	17,186	5,780	total annual exp.
INCOME					
FIXED: Annual Fees and Dues	14,280				
OPERATING: Annual Aircraft Rental Inc.	32,000	20,500	11,500		
TOTAL ANNUAL INCOME	46,280	30	NUMBER OF MEMBERS (enter)		
SURPLUS/DEFICIT (to gen. reserve)	0	476	ANNUAL FEES AND DUES PER MEMBER/YR		

ANNUAL COST PER MEMBER, PA28-201			
50 Hours	2,834	per hour:	57
50 hours commercial rental at $90/hr	4,500		

ANNUAL COST PER MEMBER, C-172			
50 Hours	1,799	per hour:	36
50 hours commercial rental at $70/hr	3,500		

Table 11.1. *Flying club and partnership comparison.*

Both the initial investment and the member's share of the fixed expenses will be far less than a partner's share of the partnership investment and fixed expenses.

Compared to renting an airplane commercially, the flying club's advantage lies in the lower hourly rental rate. This is offset by the club's annual membership dues. A club member has to fly a sufficient number of hours at the club's lower rental rate to make up the cost of the annual club dues. If you rent a Skyhawk for $20 less per hour at the club than at the FBO, but you have to pay $500 in club dues, then you have to fly at least 25 hours a year in the club to break even in comparison to commercial rental (500/20 = 25).

Legal Structure and Organization

While incorporation and the attendant transfer of liability from the members to the corporation is optional (but strongly advised) for the partnership, it is practically mandatory for the flying club. The risk of being held liable for the actions of another member in an unincorporated flying club is greater than it is in a partnership because of the greater number of members, many of whom may hardly know each other.

While the partnership can comfortably involve all partners in all decisions, the flying club has to delegate many decisions to function efficiently. It therefore has to elect a board of directors and develop a structure of officers.

And while the partnership can get by with a fairly basic agreement that leaves a lot of room for interpretation (as long as the mechanism to resolve differences is effective), the flying club needs to spell out its policies and procedures in much more detail for the benefit of its larger membership.

To effectively govern itself the flying club needs two essential documents: by-laws and operating procedures. The by-laws define the club's purpose, its organizational structure, its policies regarding all aspects of its activities, and the responsibilities of its officers. The operating procedures are the flying club's operating manual. They spell out for the membership the day-to-day operating rules, such as scheduling procedures, payment procedures, experience requirements, currency requirements, etc.

To properly structure and incorporate the flying club in terms of the legal requirements in the state where the club is located, the services of a good lawyer are indispensable. As in the case of setting up a partnership with the help of a lawyer, you will get the best results for the lowest legal fee if you are well informed, specific and detailed in your objectives, and willing to make decisions based on your lawyer's advice instead of looking to your lawyer to make the decisions for you.

Incorporation

The reason for incorporating a flying club, similar to the reason for incorporating a partnership, is to limit the liability of the individual members of the flying club. When the club is incorporated (registered with the appropriate authorities as a corporation), it becomes a separate and distinct legal entity. It enters into commitments on behalf of its

Fig. 11-2. *Archers are found in many flying clubs.*

members, and it owns the assets of the club, such as the airplanes. The members own a share in the club rather than shares of the airplanes directly. As a legally incorporated entity, the club, rather than its constituent members, is legally liable for any commitments made by it and for any consequences of the activities performed under its auspices.

If a member of an incorporated flying club causes damage or injury while operating a club aircraft, the victim's legal recourse is limited to the club, and under certain circumstances, the member causing the problem. The other club members are shielded from liability by the incorporation.

The specific procedures vary from state to state, but incorporation is an easy administrative procedure everywhere. The founding document of a corporation is its *articles of incorporation,* a standard document the provisions of which include the corporation's purpose, its right to enter into commitments (such as acquiring assets, borrowing, and contracts), its place of business, directors, share structure, and manner of dissolution. This document is the corporation's license to operate. It is distinct from the corporation's by-laws, which spell out how the corporation is structured and governed.

Organization

The flying club's governing body is its *board of directors.* Elections for the board of directors are usually held once a year, with provisions for extraordinary elections under certain circumstances. The board oversees the management of the club and makes all major decisions for the club. It is empowered to appoint *officers* who are responsible for performing the job to which they have been appointed. Given the needs of the average flying club, typical officer positions include

- president
- vice president
- secretary
- treasurer
- maintenance officer
- operations officer
- safety officer
- chief pilot
- membership officer

Not all clubs will find a need for all these functions, and being a director does not preclude someone from serving as an officer as well. Some officer positions, such as secretary and treasurer may be combined. Many smaller clubs combine the role of directors and officers. Members are nominated as directors and as officers at the same time. In larger clubs, where the membership may not be well informed about the technical skills of individual members, it makes more sense to elect the directors and let them appoint the officers after a careful evaluation of the experience and dedication of the applicants.

Bylaws

The bylaws of a flying club is its governing document. It is as important a document as a partnership agreement is for an aircraft partnership. The points addressed by the bylaws are covered below. Many elements of flying club bylaws will be familiar to you from the partnership agreements discussed in chapter 5. They address similar issues, but for a larger audience.

Purpose   The statement of purpose can be as broad or specific as you want it to be, but it should capture the spirit of your flying club. You can distinguish between general flying, touring, aerobatic flying, the objective of providing flying as inexpensively as possible, or whatever else your club endeavors to accomplish.

Meetings   This section commits the flying club to regular meetings to ensure a forum for communications. The two common types of meetings are for the general membership and for the board of directors. The general purpose and frequency of each type of meeting is spelled out, as well as provisions for calling extraordinary meetings. A big event for every flying club is its annual meeting where the accomplishments of the year and the financial condition of the club are reviewed. Many clubs also use the annual meeting as the occasion for electing the officials for the coming year.

This clause of the bylaws also addresses how decisions are made at the various meetings (simple majority, two-thirds majority, etc.). A quorum (generally defined as 50% of those eligible to participate) is usually required to make decisions.

It is also useful to specify that meetings will be conducted according to a widely accepted code of conduct, such as Roberts' Rules of Order, to provide for the resolution of issues not specifically addressed in the bylaws.

Directors This section of the bylaws specifies the number of directors the club has, the manner in which they are elected, the length of their term of office, and the scope of their powers and responsibilities. The basic idea is that they can make all the decisions necessary for the club to function, including the buying and selling of assets, committing the club to borrowing money, and enforcing FAA and club rules and regulations. The bylaws may specify that the directors seek the membership's majority approval for major decisions, such as the selection and purchase of new equipment. This may be especially desirable in a small flying club.

An important role of the directors in bigger flying clubs is to select the club's officers, usually by majority directors' votes (smaller clubs may choose to directly vote for officers). Depending on club preference as specified in the bylaws, the directors may select officers from among themselves or they may consider candidates from the membership at large.

Officers This section identifies the officers the club is to have, the manner of their selection, and the scope of their duties and responsibilities. It also addresses the issue of officer compensation. Most officers in most clubs serve with no compensation. In some larger clubs, officers fulfilling labor-intensive functions, such as treasurer, may get a number of free flying hours per month in exchange for their services.

Some functions may be combined, especially in smaller clubs. Officers are expected to attend the periodic directors' meetings and report on the status of the club under their responsibility. Typical officer positions and roles include

- President. The president is the flying club's chief executive officer; in effect, the flying club's boss. He or she is the link between the board of directors and the club, charged with implementing the board's directives. The president is also responsible for reporting to the board any concerns of the membership. An important responsibility of the president is co-signing all checks issued by the club.

- Vice president. The vice president is the president's stand in. When the president is absent, the vice president inherits all his or her powers.

- Secretary. The secretary is the flying club's record keeper. The secretary keeps the minutes of meetings, membership lists and records, and an original set of bylaws. As part of his or her corporate duties, the secretary also co-signs with the president all contracts and other legal instruments following board approval.

- Treasurer. The treasurer is the club's financial manager. He or she collects the dues, fees, and payments, pays the club's bills, keeps the books and receipts, and makes sure it all balances at the end of the day.

- The treasurer issues all checks which must be co-signed by the president. Most clubs put a dollar limit on the size of expenditures that can be paid without board approval.

- The treasurer must be chosen with great care and should be someone who has had solid experience in keeping financial records. If there is no one in the club with bookkeeping experience, have a bookkeeper set up a system, show a club member what records to keep, and have the bookkeeper come in once a month to balance the books. The bookkeeper's modest fee will be worth its weight in gold.

- Chief pilot. The chief pilot is responsible for ensuring that all members have the appropriate levels of experience and currency for the flying they do. The chief pilot is in charge of all the club instructors, if there are any, and should ideally also be an instructor. If there are no instructors in the club it is the chief pilot's responsibility to ensure the availability of qualified and reliable outside instructors for club members.

- Maintenance officer. The maintenance officer is responsible for ensuring that all aircraft are properly maintained and that maintenance takes place in a timely fashion. The maintenance officer schedules all maintenance and is also responsible for tracking squawk reports and expeditiously arranging for the performance of unscheduled maintenance.

- Operations officer. The operations officer is responsible for ensuring that the scheduling system works and for the day-to-day operations of the club. In a large flying club this function should be independent. In smaller clubs it may be combined with the maintenance officer's responsibilities.
- Safety officer. Some clubs have a separate safety officer, charged with watching over the flying club's operations and bringing issues of safety to the attention of members, officers, and directors alike. This is another post that is especially important to have in large flying clubs.

Not all positions may be appropriate for your particular flying club, so be flexible in designing your club's structure and let common sense rule.

Vacancies   This section spells out how to deal with vacancies on the board of directors or for officers that arise when someone is unable to serve out a full term. The usual solution is to have the board approve an interim successor to complete the unfinished term.

Safety board   Large flying clubs usually have a formal procedure for appointing a safety board to objectively evaluate an incident or accident involving club aircraft. Typically the safety board includes the chief pilot and the safety officer or maintenance officer, as well as another officer or two who has no role in overseeing day-to-day aircraft operations. There should be a mechanism for the pilot involved in the incident or accident to challenge the safety board's findings, usually accomplished by a hearing in front of the board of directors.

Responsibility   Some club bylaws state in a separate section what financial responsibility a member has who has caused damage to or with club equipment. The standard approach is to hold the member responsible, based on the findings of the safety board, for deductibles and whatever is not covered by insurance due to a violation of regulations. If you do not break out this section separately, be sure to incorporate it somewhere else in the bylaws.

Membership   This section specifies how many members the club is willing to accept (usually expressed as a number of members per aircraft), what qualifications applicants must have to be eligible for membership, and how new applicants are approved.

This section also defines the procedures and notice required for resigning from the club. If there are specific procedures for disposing of a member's share, such as the club's first right of refusal, to prevent the shares from being sold to someone who doesn't meet the club's membership requirements, here is the place to address the issue.

An unpleasant though infrequent task in flying clubs is the occasional need to expel a member, and this section should provide for such an eventuality. Usually a two-thirds majority vote of members is required

Disposition of a member's share upon a member's death should also be addressed here.

Member payments This section spells out what payments members are required to make for the privilege of flying with the club. It should address initiation fees, share payments if any, refundability at the termination of membership, annual dues, monthly dues, rates for aircraft, payment due dates, and delinquencies and their consequences.

It is a common practice to require joining members to pay the annual fees, as well as the first and last month's dues upon joining. This provides a financial cushion for the club in case members get into payment problems for whatever reason.

If the fleet is large and rates are likely to fluctuate, it may not be practical to list individual aircraft fees in the bylaws. In that case the bylaws should refer to an attached fee schedule, approved by the board of directors and amended from time to time.

Club finances This section defines what expenditures can be committed to on behalf of the club by the treasurer and maintenance officer without seeking specific approval from the board of directors. It also addresses any reimbursement of members for expenses incurred at the club's request, and sets a limit (usually very low) for expenditures a member can authorize for field repairs necessary to get the airplane back to home base.

Flight proficiency program and operating rules This section states the club's obligation to establish minimum requirements for qualifications, flight experience levels, minimum periodic flight time requirements, currency requirements, and checkride requirements. It also states the club's obligation to establish a set of operating rules by which the members are to conduct their flying. The actual requirements and rules are set out in the club's operating manual, which can only be amended with the approval of the board of directors.

Surplus   If the flying club is well managed it will have a running cash surplus after taking in all income and paying all expenditures. This section of the bylaws specifies how the surplus is to be used. Most clubs retain the surplus for ongoing operations as seen fit by the club per the bylaws (for example, to buy new equipment or upgrade equipment). If the club is dissolved, any surplus is typically distributed to the shareholders.

Amendments   As in the case of all legal agreements, there have to be provisions for amending the bylaws. The entire membership should vote on any amendments, and to ensure a high degree of consensus the requirement of a two-thirds majority is recommended.

Financing and Insurance Considerations

Now that you have become familiar with the flying club's structure let's take a brief look at the differences between the club and the partnership in terms of financing and insurance considerations.

Financing

The properly structured flying club is incorporated and therefore is able to borrow in its own name from a variety of sources. Since the club has no assets other than its airplanes, any prudent borrower will lend only if it can hold the airplanes as collateral.

Bank financing   Banks specializing in aircraft loans are willing to finance club aircraft on terms similar to those they extend to aircraft partnerships. A bank will lend a percentage of the value of the airplane and will secure its position by holding the airplane as collateral for the life of the loan. The bank will expect to see a reliable source of repayment in the form of an adequate level of membership dues in comparison to the club's expenses.

Depending on the lending climate, a bank may require personal guarantees from key flying club officials if the club has no track record. This is not as risky from the officials' perspective as it may seem because if they have to pay off the bank loan they can sell the airplane to repay themselves.

Private sources   An alternative to bank financing is raising money from private sources, most likely from club members who want to help the club get going. A private source has an incentive to lend to the club if the rates are attractive in relation to other investment opportunities.

The club may find the option attractive if the private source is less costly than bank borrowing. To protect the private source, the financing should be structured similar to commercial funding in terms of the amount financed in relation to the value of the airplane, and the lender should hold a lien on the aircraft financed.

Leasing   A good alternative for a club with little capital is to lease aircraft from owners who don't use them enough and would like to offset some of the cost of ownership. Flying clubs and owners have great flexibility in structuring lease agreements, including who pays what expenses. The club can pay an hourly rate for use of the airplane, some or all of the fixed expenses, or make any other agreement that suits both parties.

Flying club insurance

There is little difference between insuring a flying club and an airplane partnership, except for the insurance rates. The same concepts of hull and liability insurance apply, but because of the greater number of pilots in the flying club, insurance companies treat the club as a commercial risk and charge commercial rates which are much higher than the personal rates charged to a partnership.

In the case of flying clubs with more than one airplane, the insurance company will put the entire fleet of the club on a master policy which may offer a slight discount per aircraft compared to individual policies. Aircraft that are leased by the club may benefit from being put on the club's master policy.

Flying club members should make sure that they are named as insureds on the club's policy, otherwise they may not be covered personally in case of a mishap. The insurance company would pay the club and then attempt to collect from the member who caused the mishap. If for some reason a club is not able to list its members as named insureds, an alternative is for the club members to purchase renter's insurance.

Operations and Safety

Of all the topics to be considered in setting up and running a flying club, operations and safety are the most familiar to prospective members. Most of us have had at least some exposure to operations issues in a multipilot, multiairplane environment, and have had safety drummed into us throughout our flight training. Thus the transition to functioning under the particular operations and safety requirements of the average flying club is relatively easy.

The first step toward establishing an effective operations and safety structure for the flying club is to carefully devise a set of operating rules.

Operating rules

The operating rules spell out how the club conducts its day-to-day flying activities. Some clubs break out nonflight rules, such as billing and scheduling procedures, from flight rules. Others combine all operating rules into one document. In any event, operating rules address the following topics:

Flight line   This section lists all available aircraft, their major specifications, their equipment, and their rental rates.

Compliance with FAA and club laws and regulations   This section explicitly requires that all members must adhere to all FAA laws and regulations, as well as any club rules and procedures.

Use of aircraft   This section of the operating rules deals with all aspects of aircraft use, including but not limited to the following:

- permitted use and specific prohibitions on use, if any.
- pilot qualifications, checkride requirements, and currency requirements per aircraft.
- the pilot's obligation to meet all aircraft operating manual requirements and limitations, conduct thorough preflight checks to verify airworthiness, and report all maintenance squawks.
- rules and regulations regarding the use of instructors by club members.
- club weather and airfield minimums for operating flying club aircraft. Weather minimums may be considerably more conservative than FAA minimums. Field length minimums and runway surface restrictions may be specified.
- refueling policies and procedures.
- postflight duties regarding the aircraft, such as flight log entries, and tie-down requirements.

Scheduling   The scheduling section of the operating rules describes how to schedule aircraft and what the consequences are for not living up to the schedule either by the member or the club. This section usually addresses with whom to schedule the aircraft, how far in advance aircraft may be scheduled, and for what period of time. It also specifies cancellation policies and alternatives available to a member if the scheduled aircraft is not present at the appointed time.

Finances, payments, and billing   This section lists all annual fees, monthly dues, and aircraft rental rates in effect. It describes payment procedures, payment due dates, delinquency charges, and the consequences of uncured delinquencies.

Being grounded away from home base   This section spells out how inadvertent grounding away from home base is to be handled for mechanical, weather, and other reasons.

Lease agreement   In addition to providing each member with the operating rules, most clubs require all new members to sign a lease agreement before their first flight in a club aircraft that explicitly commits them to adhere to all terms and conditions of renting and operating club aircraft.

Putting the rules into practice

Putting the flying club's operating rules into practice requires close cooperation between the members having responsibility for operations, maintenance, and safety, and the treasurer. The day-to-day functioning of a flying club has a lot in common with an airplane partnership, but in larger flying clubs, with more than one airplane, certain areas present particular challenges. While many members may share various tasks, it is the operations officer's overall responsibility to make sure that the system works and to take decisive action to devise a solution if it breaks down.

Scheduling   A flying club confronts the same scheduling issues as a partnership, but has to provide fair access to its airplanes to a considerably larger number of pilots. To meet this need the scheduling policies of the flying club have to more explicitly address certain scheduling issues than those of a partnership.

- Minimum rental time. Typical time slots for training airplanes are two-hour blocks. Pilots of touring clubs, on the other hand, may want to be assured of having the airplane for at least half a day.
- Maximum rental time. If demand for longer-term rentals is high, it may be equally important to specify a maximum rental time slot beyond which special arrangements need to be made. A common form of control is to specify a minimum number of flight hours for which a member will be charged per day on long-term rentals

regardless of whether or not the airplane flies. A limit can be placed on the number of days an airplane may be away without special approval. Another alternative is to restrict long-term rentals to off-peak usage during the week.

- Advanced booking. Advanced booking is a great convenience, but if it is abused a limit on how far in advance a member can book may be necessary.

- Consecutive bookings. If consecutive bookings become a problem because a member ends up hogging an airplane, a limit on the number of reservations a member can make at any one time may be necessary.

- Cancellation. If a pattern of frequent last-minute cancellations emerges, a fee for canceling inside a specified cancellation time limit may become necessary.

- No shows. No shows are worse than cancellations because they show an explicit lack of caring. When you cancel you at least make the effort to let others know that the airplane will be available, whatever the reason for cancellation may be. A mandatory no-show charge can discourage this practice.

- Late returns. Late return charges should depend on the reason for the late return. As a matter of safety, there should be no penalty for mechanical or weather delays. Late returns for nonsafety reasons should carry a hefty surcharge, unless no other club members want the airplane during that period (this is easier to arrange in smaller clubs). Any late return should be immediately communicated to the club.

The scheduling policy will not work if the club doesn't have an efficient scheduling system. For clubs with no airport presence and no permanent manager, scheduling responsibilities can be handled by an answering service, rotated among volunteer members who are accessible by telephone, or via e-mail. Larger clubs with some staff presence at the airport may handle scheduling just like a commercial operator.

Flight operations In small "self-service" clubs with one or two airplanes where the scheduled members (each of whom has a key to the airplanes) simply show up at the airport and go flying, there is little scope for flight operations. Larger clubs with enough rental activity to require a presence at the airport need someone to man an operations counter and make sure that the airplanes are cleaned up and ready for action. Such responsibilities can be assigned to volunteer duty officers or paid part-time help.

For times when staff is not available, one solution for the larger club is to rent a room at the airport to which all members have a key and from which they can operate on a self-service basis, with instructions to lock the place when they are gone.

A good alternative to ensure that the airplanes get the TLC they deserve is to appoint a crew chief for each one on a rotating basis. Working in coordination with the officers of the club, the crew chief's responsibility is to ensure that everything is in top shape with his or her airplane.

Flight information flow   Just as in a partnership, the record of flying activity per airplane and member has to flow to the right places in the flying club's administrative structure for purposes of billing and maintenance scheduling. This task is easier to arrange at a larger flying club with a presence at the airport where members can fill out flight information and maintenance squawk cards and hand them in at the counter, similar to the system used by commercial operators. A maintenance squawk book should be kept in each airplane (similar to partnerships) to give each pilot an instant maintenance status update.

In small, self-service flying clubs an aircraft log, a supply of flight cards, and a lockbox in each aircraft, similar to the lockboxes used in partnerships, works well, if regularly cleaned out by a designated member who forwards the contents to the treasurer. In this case a mechanism must be devised to immediately bring maintenance squawks to the attention of the maintenance officer and the next pilot to use the airplane.

Flight proficiency and the chief pilot   All members must meet the minimum standards of competence set by the flying club. The standard of instruction, if any, offered by the club should also be uniform and high.

Many clubs set higher levels of experience than required by the FAA to check out in club equipment, especially complex airplanes (insurance companies may also require higher experience levels). For a complex, single-engine, retractable gear airplane, 200 hours of total time, 10 hours of instruction, and 25 hours in type is not uncommon. Some clubs require an instrument rating for the more powerful singles, such as the Bonanza or P-210. For twins, as much as 500 hours of total time and 100 hours of multiengine time may be required. Fairly demanding minimums may also be set for tailwheel checkouts.

Clubs may also have separate checkout requirements for different types of flying, such as night flying, mountain flying, over water flying, aerobatics, or procedures specific to a particular airport. Clubs based at airports with very long runways may require a short-field checkout before allowing members to shoehorn club airplanes into the neighborhood's smaller strips. Some particularly hairy airports may be placed entirely off limits for club aircraft.

Although the FAA requires biannual flight reviews, it is common for flying clubs to require them annually, and this is probably not a bad idea. When you fly a variety of airplanes and don't fly very often (as may be the case for quite a few club members), an annual clearing of the cobwebs makes a lot of sense. An option is to waive the annual review for club members who fly a specified minimum number of hours during the year.

As a way of maintaining the highest levels of safety, it is hard to beat currency. A big concern in flying clubs is the lengthy interval that can pass between flights for many club members. More stringent currency requirements than mandated by the FAA are a common technique employed by flying clubs to keep the membership's flying skills sharp. The three takeoff and landing requirement in the preceding 90 days to carry passengers may be shortened to 60 days and extended to include solo flight, and minimum flight hours during the period may also be required. Members who fail to stay current under the club's rules need a checkride to get back in the air. If the club has student pilots, student pilot solo currency requirements should be carefully considered.

It is the chief pilot's responsibility to ensure that these standards are sufficient and are met. In smaller clubs, the chief pilot may also have other responsibilities, but in larger clubs it is a good idea to leave the chief pilot free to concentrate entirely on piloting standards.The most effective way for the chief pilot to maintain club flying standards is to administer many of the checkrides personally, which can be done only if he or she is a certified flight instructor. If the club offers instruction, the chief pilot should check out each instructor, manage them, and set and periodically review standards of instruction with them.

But what if your club does not provide instruction and does not have a certified flight instructor among its members? Then it is best to make arrangements with an outside flight instructor to act as chief pilot to be paid independently by each member for checkrides administered. Members can certainly use other instructors to receive instruction (as permitted by the insurance policy), but to maintain uniform standards, the instructor designated as chief pilot should be used to give periodic recurrent checkrides and to check out members in club equipment.

Flying club instructors Flying clubs handle their association with flight instructors in a variety of ways. Some have members who are instructors, others may keep instructors on staff, and still others maintain a list of approved outside instructors with whom club members make arrangements on their own.

There are no hard and fast rules about which option is best, but it is fair to say that having instructors contract their services as independent contractors directly with members has some advantages.

Safety

Preventive measures are the most effective tool for ensuring a high level of safety in the flying club's operations. Good training, experience, and currency are the prerequisites of safe flying.

FAA required minimum experience levels are just that–minimums. Prudent flying clubs often demand more. The same goes for operational minimums, such as VFR visibility and ceiling minimums and IFR approach minimums. Overseen by the chief pilot or the safety officer, the flying club should carefully evaluate these factors in view of the members' experience and the kind of flying the club does and set standards the club deems appropriate.

The practice of charging a monthly flight time minimum regardless of whether or not a member flies has an effect on safety in addition to making financial management more predictable by providing an incentive to fly regularly. Another common technique to heighten safety awareness is to require the filing of a flight plan for any flight more than 50 miles away from home base. Having members regularly attend the FAA's safety seminars and arranging in-house seminars tailored specifically to the club's circumstances are also useful tools to maintain and improve flying safety. And if a club is large enough, appointing an independent safety officer may also be warranted.

Maintenance

How complicated a task maintenance is for the flying club is a function of club size. The greater the number of members and aircraft, the more formal maintenance procedures need to be.

Maintenance is a big concern for potential members, many of whom may be wary of budget maintenance. To allay their fears it is best for any flying club to have a stated maintenance policy and strictly adhere to it.

Maintenance policy   The maintenance policy is a statement by the flying club of its maintenance philosophy and objectives. The objective of providing quality maintenance is a given, but how this is accomplished can vary greatly depending on the flying club's purpose and resources.

A training and pleasure flying club formed specifically to provide inexpensive flying for members on tight budgets may want to state explicitly in its maintenance policy that quality maintenance is to be provided at the lowest possible cost. It may also want to specify that members under the supervision of a mechanic are to perform as much maintenance as possible.

A flying club of high-priced IFR touring aircraft flown primarily on business by pilots with relatively deep pockets might have a different approach to maintenance. Its policy may state that in the normal course of the club's business maintenance problems are to be fixed under a maintenance contract with an FBO as expeditiously as possible and may specify that members are not to work on the airplane.

Very large flying clubs may have an in-house maintenance department and a policy describing its duties and responsibilities.

Maintenance officer   It is a good idea for any flying group of six or more members to designate one of the members as the maintenance officer. The maintenance officer's responsibility is to establish and oversee a system that gets all of the group's maintenance needs accomplished in a timely fashion. Typical responsibilities include tracking and arranging the resolution of squawks, tracking and setting up scheduled maintenance, and being the primary point of contact on behalf of the group with the mechanics performing the work.

Sources of maintenance   The sources of maintenance are no different for a flying club than for a partnership. Maintenance may be arranged through part-time mechanics, FBOs specializing in light aircraft, and large FBOs that maintain complex aircraft including executive jets. The flying club's choice is a matter of maintenance policy and available budget.

Case Study: Extra 200L–Eight Partners

The aircraft

The Extra 200L is a 200-hp, low wing, high-performance aerobatic airplane made in Germany. Its benign flying characteristics make it the only production aerobatic airplane that can be used for basic aerobatic training all the way through unlimited aerobatic competition.

The nature of aerobatic flying lends itself to large partnerships. Aerobatic sorties are short and exhausting. Few pilots can withstand more than three brief sorties a day, and one to two is more common.

Case 11-1. *None of the eight partners in this Extra 200 could afford to go it alone.*

The Extra 200L is an expensive airplane. A large partnership is a convenient way for pilots to have access to the Extra 200L who could otherwise never afford the quality of aerobatics offered by the airplane.

Partnership background

This is an interesting partnership in that the instigator behind it is an aerobatic school. The school's operator saw the partnership as an excellent opportunity to provide aerobatic students, who were ready to move into their own airplane, with the chance to remain associated with the school and own a share of a top aerobatic aircraft. The aerobatic school manages the aircraft for the partnership, removing the burden of operational chores from the partners. The school receives a management fee for its services, which is a small percentage of the billings.

Partnership structure

The partnership is incorporated, which is the only reasonable form of organization for a group as large as this engaging in an activity such as aerobatics. The partners own equal shares of the aircraft. For all practical purposes this partnership is really a small flying club. When a partner wishes to sell his share, the existing partners have the right of first refusal, and new partners must agree to accept the existing rules and conditions of the partnership.

The by-laws specifically spell out the experience and currency requirements for flying the airplane in view of its use for aerobatics. Shareholders must be checked out by the aerobatic school that manages the airplane and must meet the school's currency requirements. The checkout is a rigorous affair and authorizes the pilot who passes it to fly only the maneuvers for which he or she was checked out.

With this many partners, insurance is at the commercial rate. Incorporation and the commercial insurance coverage provide each partner adequate liability protection.

Finances

The partnership's fixed expenses are shared equally. Insurance is paid in one lump sum annually. Other fixed expenses are billed monthly. Operating expenses are paid in full by each partner per the number of hours flown by the partner. Operating expenses include an engine overhaul reserve, a propeller overhaul reserve, and a general maintenance reserve. Each partner buys his or her own fuel directly from the FBO, independent of the partnership.

The partnership breaks down its finances as follows:

ANNUAL FIXED EXPENSES ($)		(8 Partners)
	Total	Per Partner
Hangar	4,800	600
Insurance	1,400	175
Annual	768	96
Parachute Repacking	192	24
Total Fixed Expenses	7,160	895
		(75/mo)

HOURLY OPERATING EXPENSES ($)

Fuel	30
Engine Reserve	8
Propeller Reserve	2
General Maint Res	10
Total Hourly Op. Exp.	50

TOTAL EXPENSES PER YEAR ($)

50 hours of Flying Time	3,395
Hourly Rate @ 50 hr/yr	68

An Extra 200L for $68 per hour! That is the cash expense, because the noncash cost of capital of each partner's share is not considered by the partnership. One one-eighth share in the Extra costs $28,500. If we assume an annual cost of capital of 7%, that works out to an additional cost of $1995 for total annual expenses of $5390, which is $108 per hour, still a bargain for a brand new airplane that costs $228,000.

This partnership illustrates the financial benefit of having a large partnership. The large number of partners dramatically reduces the per partner share of the high annual fixed expenses ($75 per month per partner) and makes the relatively high hourly operating expenses bearable.

Operations

Operations are very straightforward from the partners' perspective because everything is taken care of by the aerobatic school that manages the airplane for the partnership.

The airplane is scheduled through the school's reservation desk on a first come, first served basis. A partner is not allowed to schedule more than one overnight trip every two months without prior approval from the other shareholders except to attend aerobatic competitions, which take precedent over all other scheduling. In practice several of the partners are likely to attend the same competitions, sharing the aircraft (an aerobatic competition sequence is about 7 minutes in duration). Attending competitions as a group is a highlight of the partnership experience.

Partners receive instruction from the aerobatic school that manages the airplane at a discounted rate.

Maintenance

Maintenance is performed by the aerobatic school that manages the airplane. The cost schedule is designed to accurately reserve for maintenance needs and may be periodically adjusted as required. If the maintenance reserve proves inadequate, members are assessed for their share of the extra expense based on their percentage share of the hours flown. A similar clause covers the cost of overhauling the engine if it is required before TBO and the engine reserve proves insufficient.

The partnership experience

This partnership is a recent one so it is too early to assess its long-term success. It is certainly one of the more innovative ones, and a very important development in cooperative flying. If we are to succeed in revitalizing general aviation we need to make exciting, quality flying affordable for a much greater number of pilots and potential pilots. This partnership is one of the more inspirational steps in that direction.

12

Buying the Aircraft

When you have a good idea of the kind of flying you want to do and how much money you have available, you need to decide on the aircraft make and model that will best meet your performance requirements within your budget, and you have to find a specific aircraft to buy. Comprehensive information and good planning are essential to move quickly and smoothly from your initial thoughts of make and model to flying away with a freshly signed bill of sale and the just completed aircraft registration naming you and your partners as the co-owners.

You need to know how to find detailed specifications and performance information on the greatest number of makes and models that meet your objectives. You need to find reliable pricing information. You need to line up financing and insurance as far in advance as possible to avoid delays once you have located a specific aircraft. You have to identify aircraft for sale worth looking at, and after all that work you should really buy one. This chapter shows you how to conduct the initial research into aircraft specifications and performance, how to develop pricing information, and how to find aircraft for sale worth investigating.

Researching Aircraft Specifications

Chances are that a wide variety of aircraft types will fit your mission requirements. Whether your primary objective is training, touring, or even aerobatics, there is a vast private air force for sale out there ready to meet your needs. How to be aware of all the options, and how to choose? As you conduct a systematic, carefully researched comparison of all the options, you will suddenly reach a point where the ultimate choice becomes obvious.

Pilots are a conservative and loyal lot. They tend to stick with what they know. If they learned to fly on Cessnas they will most likely buy a Cessna. If they have been renting Cherokees from the local FBO, that is what they will want to buy. Yet there may be other alternatives worth considering. In recent years, for example, many pilots purchased used Piper Arrow IIIs and IVs, unaware that a used Grumman/Gulfstream Tiger was an excellent and considerably less expensive alternative. The price was lower because except for one brief interlude the Tiger has been out of production for some time, raising spare part questions, it was not well known, especially by pilots who came on the scene after its production had ceased, and it had an undeserved reputation of being a "hot ship." Yet if your performance and specifications research had identified the Tiger as an alternative worth investigating, you would have been pleasantly surprised when you looked into prices to find the option of Arrow performance for up to $10,000 less.

Being aware of all the aircraft types available to meet your needs based on comparative performance is only one aspect to research. For each type you consider, you should also find out how well it has done in service, what ADs and service bulletins have been issued on it, and what quirks if any it is known for.

Technical specifications to consider

The technical specifications you will find most relevant will depend on your intended use of the aircraft, but there are a number of standard specifications you should look for when you develop the information to make comparisons. Specifications such as dimensions, which are nice to know but are of little help in comparing types, are not presented here.

- Aircraft make, model, year
- Seats and configuration
- Power plant make, model, year
- Horsepower
- Type of fuel used (80, 100LL, auto, etc.)
- Propeller make and model
- Weights and loadings

- Gross weight
- Empty weight
- Useful load
- Payload, full fuel and oil
- Fuel capacity
- Baggage capacity
- Performance
- Maximum speed, sea level
- Cruise speed (75% power, 65% power, 55% power)
- Range (75% power, 65% power, 55% power)
- Rate of climb, sea level
- Service ceiling
- Best angle of climb speed
- Best rate of climb speed
- Stall speed, clean
- Stall speed gear and flaps down
- Approach speed
- Takeoff distance over 50-foot obstacle
- Landing distance over 50-foot obstacle

Sources of information

Technical information on aircraft is available from a variety of sources. Some are legendary tomes, such as *Jane's All the World's Aircraft* (updated annually) or, for older aircraft, *Juptners* many volumes. They are prohibitively expensive for infrequent use, but any good library should be able to track them down for you. (Jane's occasionally puts out condensed paperback extracts, but these are not sufficiently detailed to be of use to purchasers of aircraft.) To meet the needs of most of us, there are other more readily available sources.

Books There are several good books on the market which are compilations of information on general aviation aircraft. These books group aircraft by class (single-engine, multiengine) and type within class (two-seat trainer, four-seat fixed gear, etc.). They provide brief histories as well as all the important technical specifications.

Aviation Consumer This publication is perhaps the most comprehensive source for not only technical information on an airplane types but especially on how well each has done in service. Technical reliability, maintenance problems, safety record, major ADs and service bulletins, and handling are all addressed in comprehensive pilot reports. The publication accepts no advertising, which puts conspiratorial minds at ease. Most production aircraft have been covered at some time or other, and many are covered again and again periodically. McGraw-Hill has issued a large collection of these aviation consumer reports in book form.

Aviation magazines   Aviation magazines are in the business of writing pilot reports on practically every type of general aviation aircraft from antique Piper Cubs to hefty brand new twins. The latest offerings of airplane manufacturers and used aircraft are given equal time. The more popular used types are reviewed every 3 to 4 years. A good strategy is to get copies of all the articles over time on a particular type. Periodically magazines also publish articles comparing the competing models of several manufacturers. Some magazines provide copies of pilot reports for a fee. Many publish indexes from time to time. If you have difficulty finding back issues, your librarian can help. Although most aviation magazines are too specialized to be in most libraries, the librarian can research which library closest to you has back issues.

Aircraft manuals   It is relatively easy to borrow the manuals of the aircraft in which you are seriously interested. Many manuals are also available for purchase, but it makes sense to borrow them.

Airworthiness directives   During your initial research you can familiarize yourself with the most important ADs from book and magazine articles. But such secondary sources of information are inadequate once you settle on the particular type of aircraft you expect to buy. Books and magazines are dated. ADs may have been issued after their publication or they may cover only the major ADs.

To be fully informed of all ADs issued on an aircraft type, you need to order an official list of ADs. This list can be obtained from your local GADO, the FAA in Oklahoma City, or through AOPA which provides a special AD list service for a modest fee. For an additional fee AOPA will also provide all FAA Seville difficulty reports.

It is crucial that you get an official, up-to-date AD list for the type of aircraft you decide to buy. Comparing the AD list to the aircraft logs is the only effective means of independently verifying compliance with all ADs of the aircraft you will buy.

Finding and Pricing the Aircraft

Developing a shortlist of aircraft types based on technical specifications, performance data, and pilot reports is only half the battle. You also have to develop reliable price information, and you have to start looking for specific aircraft for sale.

Aircraft advertised for sale

There are several publications dedicated wholly to advertising aircraft and related products for sale. The most well known and comprehensive among them is *Trade A Plane*. Its yellow, newspaper size pages carry literally thousands of aircraft advertisements in each issue. *Trade A Plane* is reasonably priced, appears three times a month, and is available by subscription for periods as short as 3 months, the ideal "window" for most private buyers. The ads generally contain sufficient information to group the offered airplanes according to your specifications and will enable you to develop a good idea of price ranges.

Other good sources of airplane ads are regional aviation papers such as the *Atlantic Flyer* and *Pacific Flyer*, which carry large classified sections. For homebuilts, the classified sections in *Kitplanes* and EAA's *Sport Aviation* magazine are helpful. *Soaring* magazine lists gliders for sale.

There are a number of glossy ad magazines, such as the *A/C Flyer*, with pages and pages of pretty color pictures of aircraft for sale. These publications list aircraft offered by dealers, and cover mostly general aviation business aircraft, but they also carry some single-engine light aircraft, especially the more expensive ones such as Mooneys and Bonanzas. Prices are usually at the high end due to hefty dealer markups.

Computer listings are a relatively recent source of aircraft advertisements. Most list aircraft for sale by owners (no dealers). The lists themselves are advertised in the classified sections of most popular aviation magazines.

Aircraft Blue Book Price Digest

The best source of current price information is the *Aircraft Blue Book Price Digest*. However, it is only available to aviation businesses such as dealers, aircraft financiers, and insurance companies. It is also very expensive. The *Aircraft Blue Book* appears four times a year. It tracks in great detail the wholesale and retail prices of virtually every aircraft make and model produced, adjusted for engine time, equipment, and general condition. It is not easy to read. There is a complex adjustment process to the basic prices, and the prices are in code. This is because the blue book is intended to benefit business users (especially aircraft dealers) who pay the hefty subscription price.

In spite of all the secretiveness, there are informal ways for private individuals to obtain blue book information. Some subscribers may see a business reason to reveal retail prices to you. Contacts with personnel working at firms which subscribe may lead to a willingness to provide counsel on aircraft prices implicitly understood to be based on blue book values. Information can also be obtained from your friendly banker. Ask what maximum percentage of the going price the bank is willing to finance, and is the price wholesale or retail. Then submit a list of aircraft types with desired avionics and engine and airframe times and ask the banker for the maximum dollar amount he or she would be willing to finance on each type. You can then figure out the approximate blue book price.

Other sources of information

Other sources you should cover in your search are the traditional ones, such as word of mouth, the region's FBOs, and bulletin boards at airports. Bulletin boards can be an especially helpful source of information on pilots seeking partners. A twist on the bulletin board approach is to put up your own buyer's ad.

Financing and Insurance Homework

It is advisable to line up financing and insurance as early as you can to minimize delays when you find the aircraft you want to buy. Once you decide to buy a particular aircraft you will most likely be required to put down a deposit which will hold the aircraft for a specified period of time while you complete the necessary arrangements. If at this stage you run into delays through no fault of the seller, you may risk losing your deposit.

The best time to start working on financing and insurance is when you have made a decision on the type of aircraft you want to buy and have defined such criteria as total time, engine time since major overhaul, and avionics requirements.

Financing preapproval

Most banks are willing to prequalify you for a certain amount of financing based on predetermined criteria for the aircraft. You fill out all the loan applications and provide all the other information the bank needs, and you define in great detail the specifications of the aircraft you are looking for. The banker can then preapprove a loan for you which will be immediately available when you locate the aircraft that fits the predetermined criteria and complete a title search.

Banks can usually approve an aircraft loan in 48 hours, so why the need for preapproval? You need it to find out if you qualify for a loan. Some pilots make the mistake of not getting preapproval because of the fast turnaround in bank approvals, only to find out that they don't qualify for a loan on the terms they need. There is then a mad scramble for alternative financing, which may or may not work out in time, threatening loss of the deposit given to the seller.

Lining up insurance

Insurance can be lined up much in the same manner as bank financing. The drawback of not lining up insurance ahead of time is not that you will not get coverage, but that coverage will be on less favorable terms than you would like. Determine the levels and types of coverage you want, define the specifications of the aircraft you are looking for, ask how much coverage will cost, and have your insurance provider standing by on your terms and conditions.

When you have completed your research of aircraft specifications, performance, and prices and have made a decision on what to look for, it is time to see what's out there. Again, a systematic approach will yield the best results. Draw up a list of likely airplanes from all the sources available to you, and start making some phone calls.

Beginning the Search

The geographic location of the airplanes for sale is an important question when deciding which ones to check out. To have the widest choice available it is best to be as flexible as possible in how far you are willing to travel to inspect an airplane. A good rule of thumb is to have the airplane brought to you if it is within reasonable flying distance from your home airport. You have the advantage of home turf for the test flight and your mechanic can do a prepurchase inspection. However, if the airplane is far away, or demand for it is great, you and your mechanic may have to go to it.

Define what expense you consider reasonable for such travel, consider it part of the cost of buying an airplane, and set your geographic limits accordingly. Sometimes it is possible to make arrangements with the seller to fly to your home field for preagreed expenses should you decide not to buy the airplane following the prepurchase inspection. An inexpensive but somewhat limiting solution is to restrict yourself to checking out airplanes only within day trip range.

Initial Contact with the Seller

A lot can be learned from the first telephone call. If you are well organized and ask the right questions you can quickly screen your initial list and decide which aircraft merit further follow-up. You might want to fill out a screening worksheet for each aircraft and keep a record of the responses to your questions. The questions below are in appendix 6 in worksheet form suitable for reproduction.

- How many owners has the airplane had?
- Where has it been based geographically during its life?
- Has it ever been used for training or rental flying?
- What is the total time on the airframe?
- What is the total time on the engine?
- What is the total time on the engine since major overhaul?
- How many times has the engine been overhauled?
- Who performed the overhaul? The choices are the factory (the best but most expensive alternative), an FAA certified repair station (many specialize in engine overhauls and are perhaps the best buy for the money), or an FBO or independent mechanic (who do some of the best and some of the worst work around; the problem is knowing who does the best and who does the worst).
- Was the overhaul to factory new tolerances or service limits? Factory new tolerances mean that the main components meet the standards of new components. Service limits mean that wear and tear is within the limits within which the components may be kept in service, but the components may be considerably below factory new tolerances.

- Were the accessories also overhauled? If not, how many hours are on the accessories (starter, alternator, magnetos, vacuum pump, etc.)?
- Are all ADs complied with and entered into the aircraft logs?
- Has there been a top overhaul since the major overhaul?
- Is there any damage history?
- How old is the paint and interior?
- Rate the exterior and interior on a scale of 1 to 10.
- Has the airplane been hangared?
- When was the last annual inspection?
- Who did the last annual inspection?
- Is the oil sent out for analysis at oil change? Are the results available?
- How many hours has the airplane flown since the last annual inspection?
- How many hours has the airplane flown in the last 12 months?
- When was the most recent transponder and static/altimeter check?
- What major maintenance items have there been in the last 12 months?
- What make and model avionics does the airplane have (coms, navs, HSI, ADF, GPS, etc.)?
- Are there any maintenance issues with the avionics?
- Does the airplane have an intercom? Does it work?
- Are the airframe and engine logs complete?
- Are pictures available (exterior, interior, and instrument panel)?
- What is the asking price?
- Does the airplane have EGT and CHT gauges?

Which airplanes you decide to follow up on depends on the standards of acceptability you have set for yourself. These may include limits on total time, time since engine overhaul, acceptance of engine overhauls done to factory new tolerances by at least an FAA authorized repair station, no use of the airplane as a trainer or rental aircraft, limits on geographic location during the life of the airplane (you may not, for example, want to consider an airplane that has spent the last 10 years unhangared in a saltwater environment), and ease of getting together with the seller for the prepurchase inspection.

The Prepurchase Inspection

A thorough prepurchase inspection is absolutely imperative prior to buying an aircraft. If you do not have substantial aircraft maintenance experience it is also imperative that an A&P mechanic of your choice perform the inspection. The small inspection fee is well worth the peace of mind it will bring, considering the size of your investment. The seller will often reassure you that an annual inspection has just recently been done, or that the seller's mechanic will be happy to give an opinion. Most sellers are honest and are just trying to be helpful, but you can never be sure unless you use your own mechanic.

A prepurchase inspection should usually consist of three phases: examination of the aircraft papers, the mechanical inspection of the airplane, and the flight test. Below are some guidelines for topics to cover in each phase. Use them only as guidelines and rely on your mechanic for best results. As it has been said since Roman times, "Let the buyer beware."

Examining the aircraft papers

It is a good idea to start with the aircraft papers. If there is a problem with them you can save everyone a lot of time by discovering it up

Fig. 12-1. *A prepurchase inspection away from the current airplane ownership.*

front. The review of the papers is also important simply to decide if you feel comfortable going on the test flight. You should concentrate on the airframe and engine logs, but you should also examine all the papers required to be on board by the FAA:

- certificate of airworthiness registration
- In the airframe and engine logs you should check for
- proper annual entries
- evidence of 100-hour inspections indicating commercial use
- compression (most recent and history)
- number of hours flown per year
- evidence of unscheduled repair work
- airworthiness directive compliance
- major engine overhaul (when and where done, to what tolerances)
- record of geographic movements

The mechanical inspection

The mechanical inspection should be along the lines of a 100-hour inspection, following the manufacturer's 100-hour inspection worksheet. An especially important part of the inspection is the compression check. If the engine has been overhauled, be sure to verify overhaul information provided by the seller on initial contact (when was the overhaul done? by whom? to what tolerances?). The following guidelines are provided to familiarize you with the items your mechanic should be checking during the prepurchase inspection.

Airframe

- Check for wrinkled skin, loose rivets, dings, cracks, and corrosion.
- Check for mismatched paint, which could be a sign of repairs.
- Check all controls for free and correct movement.
- Check all control hinges (ailerons, elevator, rudder, flaps) for looseness, play, and hairline cracks.
- Check vertical and horizontal stabilizer attachment points for looseness, play, and hairline cracks.
- Check wing attachment points for hairline cracks and corrosion (there better not be any looseness).

Fig. 12-2. *Have a maintenance specialist in a particular type check it out for you.*

- Check control cables for looseness and chafing. Look inside fuselage and wings through inspection panels with a flashlight.
- Check fuel caps, quick drains, and fuel tank areas for signs of fuel leaks (brownish stains).
- Check fuselage underside for cleanliness and signs of leaks from the engine area.
- Check wing struts for any signs of damage, corrosion, or hairline cracks.
- On fabric covered aircraft, do fabric test and check for loose or peeling fabric.
- Check engine cowling for looseness, play, and cracks, especially at attachment points.

Landing gear

- Check landing gear struts for leaks.
- Check tires for wear and bald spots.
- Check brake pads for wear, brake disks for corrosion, pitting, and warping, and brake hydraulic lines for signs of seeping or leaking fluid.

Cockpit

- Check cabin doors and windows for signs of water leaks.
- Check windows for crazing.
- Check seat belts for wear and tear. They can get caught and damaged in the seat rails.
- Move all controls and trim to verify full control movement and check for binding.
- Move all other knobs and switches to check for proper operation.
- Check ELT for proper operation.
- Check entire aircraft for proper display of placards and limitations.

Engine and propeller

- Check compression.
- Check baffles for damage or deformation. Baffle irregularities can cause cooling problems.
- Check for any sign of leaks, especially around the various gaskets. Look for oil and fuel stains.
- Check lower spark plugs for proper condition (take them out and examine them).
- Check wiring harness for signs of brittleness and fraying.
- Check the induction/exhaust system for leaks, cracks, corrosion, and looseness.
- Check engine controls running to the cockpit for free and easy movement.
- Check the battery fluid level and for signs of overheating.
- Check accessory attachments (alternator, magnetos, vacuum pump, starter, electric fuel pump, etc.). Check alternator belt for fraying and tightness.
- Check propeller for nicks and spinner for cracks.

The test flight

If the paperwork and mechanical inspection check out, you are ready for the test flight. Insist on one, and offer to pay for the gas if necessary. Some sellers are wary of phony purchasers out to get some free flying.

The objective of the test flight is twofold. It is your way of verifying that the airplane works as advertised, and it is also your chance to see if you feel comfortable with the way the airplane handles, if it is really the type of airplane you want. Handle the test flight systematically and professionally. For pilots inexperienced with test flights, a good way to go about them is to treat them as a VFR flight test. Such a test requires the pilot to operate the aircraft through the whole spectrum of its normal operating range and is as much a mechanical test of the airplane as it is of the pilot's abilities. There won't be an examiner prompting you to do this and that, so take a checklist of all the items you want to accomplish with you. With that in mind, consider the guidelines below.

- Preflight. Do a detailed preflight per the operating manual. Follow the detailed manual checklist to be sure of covering everything.
- Engine start. Note how easily the engine cranks and turns over.
- Taxi. Perform radio checks on both coms. Test brakes. Notice engine response, steering, and suspension.
- Pretakeoff checks. Perform careful pretakeoff checks and run-up, verifying aircraft behavior to operating manual standards.
- Takeoff. Note if power settings reach required levels. Climb to a convenient maneuvering altitude.

Fig. 12-3. *Fly before you buy.*

- Basic VFR maneuvers, handling checks. Perform steep turns, slow flight, and stalls. Cycle flaps and gear. Note if aircraft is in trim.

- Cruise checks. Set up various cruise configurations. Note performance compared to operating manual standards. Check engine gauges.

- Avionics checks. Check all avionics in flight. Do in-flight radio checks. Track VORs, localizer, and glide slope on all appropriate navigation charts. Compare needle indication of each VOR set to the same radial. Check ADF and marker beacons as appropriate. Establish radar contact with ground control, compare transponder altitude readout reported by ground control to the altimeter indication.

- Miscellaneous checks. Check cabin lights, vents, and heating system.

- Touch and goes. Perform two or three touch and goes to feel fully comfortable with the handling of the aircraft.

- Postflight check. Uncowl the engine, check for evidence of any leaks. Turn on all navigation lights, landing lights, strobes, and flashing beacon and check for operation.

Decision Time

If the test flight goes as well as the paperwork review and the mechanical inspection, and the price is what you expected, it is decision time. Review carefully the results of the prepurchase inspection. See where there may be grounds for a price adjustment. Propose to the seller any price adjustments you feel are appropriate or items to be taken care of on the airplane before it would be acceptable to you. If you can reach agreement it may be time to give the seller a deposit and move on to closing the deal.

You have found the airplane you want, and you and the seller are in agreement on the price, so what next? There is no need to panic. As with a car or a house, there are some simple but essential steps to be taken to complete the transaction and take possession. You need to do a title search to find out if the seller really owns the airplane free and clear of any liens. You need to complete your financing arrangements, you need to get insurance, and you have to set up the FAA paperwork to transfer title. During this time the seller has to complete whatever additional repairs or inspections you may have agreed to.

The Deposit and Purchase and Sale Agreement

Both you and the seller need to see some commitment to your pending deal. This is accomplished by your giving a deposit to the seller and both of you signing a purchase and sale agreement. The deposit is evidence of your commitment. It holds the airplane for you for a set period agreed to by you and the seller. If you do not buy the airplane during the time the deposit is in effect through no fault of the seller, you lose the deposit. The amount of time for which a deposit will hold an airplane is typically 30 days.

The deposit need not be a big amount in relation to the airplane's price, but it should be big enough from the seller's point of view to keep you motivated to complete the transaction. About 2% of the airplane's price is usually sufficient. Thus on a $50,000 airplane, a deposit of $1000 would be appropriate.

The purchase and sale agreement documents the terms under which you and the seller agreed to the sale. It is the seller's commitment not to sell the airplane while the deposit is in effect. It spells out the purchase price, the terms of the deposit, and any other conditions you may mutually agree to. If you show up with the money before your deposit period expires, the seller has to sell you the airplane. If he or she reneges, you may have recourse in the courts. The purchase and sale agreement also gives you protection if you set conditions which must be met by the seller (such as fixing mechanical problems or having an annual inspection done) before the deposit period expires. If the seller does not meet these conditions, you can back out and your deposit must be returned. Be aware, however, that once the agreement is signed, you cannot come up with additional conditions not covered in the agreement.

Purchase and sale agreements should be tailored to meet your personal circumstances and should be reviewed by your lawyer. The agreement should be signed by both buyer and seller and should be dated. At a minimum the following information should be covered:

- aircraft make, model, serial number, and N number
- detailed equipment description
- sale price
- amount of deposit
- amount of time for which the deposit holds the aircraft

- satisfactory title search
- aircraft to be delivered in full working condition as it was at the time of the prepurchase agreement, subject to any special agreement and noted discrepancies addressed in the agreement
- special conditions (examples are annual inspection by the seller, repair by the seller of mechanical problems identified during the prepurchase agreement, a new paint job, etc.)

The escrow option

Aircraft purchasers have the option of making sure that the seller does not disappear into the sunset with their cashed deposit, never to be seen again, or spending the deposit money and not being able to return it when there are grounds for its return. This option is the escrow account, a practice popular in real estate and other business transactions. It is an account maintained by an impartial third party, usually a lawyer. The money is no longer under your control, but is not released to the seller until the deposit period expires and you fail to buy the airplane, thus entitling the seller to the deposit.

The Paperwork

The big job after making a deposit on an airplane is getting all the paperwork lined up to transact the sale. A most important piece of paperwork is the title search.

The title search

A title is proof of ownership of an asset. In the case of airplanes, a title is filed with the FAA in Oklahoma City. The owner of the asset can pledge it as collateral against debt, giving the lender a right to the asset (a security interest). If the debt is not properly handled, the lender can take possession of the asset and sell it to extinguish the debt. Sale of the asset does not invalidate the lien. Thus the new owner can have the asset snatched from him by the lien-holding creditor of the former owner without being able to do anything about it. In fact, the lender's agreement with the former owner probably forbids the sale of the aircraft without extinguishing the loan, so the buyer can also find himself the victim of a fraudulent sale. Thus it is essential that the buyer perform a title search to verify that title is clear or has only the lien on it known to the buyer–the underlying debt to be paid off from the proceeds of the sale in exchange for release of the lien.

Title can be further complicated by the ability of creditors to put a lien on a debtor's asset even if it is not pledged to the creditor. An example is a mechanic's lien for unpaid maintenance. If the case goes to court and the lienholder wins, he can force the sale of the asset regardless of who the new owner is. A buyer would be insane not to run a title search before buying an airplane.The easiest way to do a title search is to use AOPA's title search service. If you are getting bank financing for the airplane, the bank will do the title search.

What if the title search shows a lien unknown to you? It may be a deal breaker, but more often it is an old lien that was not released upon payment of the underlying debt. The lienholder has to specifically request that the lien be removed, and many lienholders never bother to do this. They then have to be contacted and asked to have the lien removed. If you encounter any liens, anticipated or unanticipated, make their removal a condition of purchase.

The option of title insurance

It can happen that a routine title search does not reveal a lien which could cause problems if it surfaces. Such occurrences are extremely rare, but to put your mind completely at ease about any title problems, you can purchase title insurance. Ask your insurance provider for advice.

Completing finance and insurance arrangements

Upon placing a deposit on an aircraft you should complete the financial arrangements if you are obtaining bank financing, and you should also arrange insurance.

If you have preapproval, the financing arrangements will be closed quickly. You have to provide the bank with all the details of the aircraft and they will run a title search and prepare the loan documents. When everything is ready, you have to sign the loan documents and are given a bank check made out to you and the seller, which you will both have to sign before it can be cashed.

Insurance arrangements can also be quickly completed. You have to provide the insurance company with the particulars of the aircraft and pay for the policy. In return the insurance company will issue you a binder, a brief, temporary proof of immediate coverage pending the arrival of your policy later in the mail. If you are getting bank financing, the insurance binder must be delivered directly to the bank by the insurance company as a condition of disbursing the loan.

The bill of sale

This is the form that will transfer title to you from the seller. It is legal proof of your purchase of the aircraft. It is usually provided by the seller, but it doesn't hurt for you to have a blank form with you just in case. The form can be obtained from any General Aviation District Office, or from the FAA's headquarters in Oklahoma City (AC Form 8050-2). There are instructions on the form on how to fill it out and where to send it.

Registration

When the aircraft is sold, it has to be reregistered in the name of the new owners. This is done on the Aircraft Registration Application (AC Form 8050-1). You must have this form with you to take possession of the aircraft. The pink copy of this form is your temporary registration until the processed permanent registration is sent to you. This form also has instructions on how to fill it out and where to send it.

Certified check for your equity portion

Most sellers will want payment from you in the form of a certified check. Be sure to have it prepared in the right amount and the right name. Have it made out in both your names and endorse it at the

Fig. 12-4. *A pretty cockpit, but don't let that be the deciding factor in buying an airplane.*

closing. It is a safety precaution. If the check is not endorsed by you, it cannot be cashed. If you are getting bank financing, your check will have to be made out for the difference between the purchase price and the bank check minus the deposit you already gave.

Taking Possession

The actual closing of the transaction is the easiest part of buying the airplane. Be sure to have all the documents listed above with you. Take one more close look at the aircraft, especially if the seller has been flying it since you agreed to buy it. Review carefully once more the airframe and engine logs. Check to see that all the conditions laid out in the purchase and sale agreement have been met. Then it is time to make it happen.

You should receive the following paperwork from the seller:

- signed aircraft bill of sale
- airworthiness certificate
- airframe and engine logs
- aircraft operating manual
- weight and balance certificates

    You give the seller

- funds in the amount of the sales price (minus the deposit already given).
- a copy of the bill of sale for the seller's records. This is optional but most sellers will want one (you will have to make a copy because the FAA form is only in duplicate, the original to be sent in, the copy to be kept by you).

    The following paperwork should be placed in the aircraft to meet FAA regulations:

- airworthiness certificate
- new temporary registration
- operating manual
- weight and balance certificates

    And now it is all done. Fire it up and fly away! If all partners are equally qualified and available, perhaps you will want to draw straws to see who gets the honor of being the pilot in command on the first flight. Just be sure to mail at your earliest convenience the bill of sale and the registration application along with all the appropriate fees.

Appendix 1

Financial Analysis

AIRCRAFT COST WORKSHEET

Aircraft Type: [   ]

(Calculate total cost in first column followed by per pilot costs for selected number of pilots)

NUMBER OF PILOTS

ANNUAL FIXED EXPENSES

Tiedown/Hangar

Insurance

State Fees

Annual

Maintenance

Loan Payments

Cost of Capital (non-cash)

Total Fixed Expenses / yr

HOURLY OPERATING EXPENSES

Fuel

Oil

Engine Reserve

General Maint Res

Total Op Exp / hr

TOTAL HOURLY EXPENSES: (annual fixed exp/hrs flown)+hourly operating exp

50 Hours

100 Hours

150 Hours

Hourly Commercial Rental

TOTAL ANNUAL EXPENSES: total hourly expenses x hours flown)

50 Hours, Own

Own (cash only)*

Rent

100 Hours, Own

Own (cash only)*

Rent

150 Hours, Own

Own (cash only)*

Rent

(*subtract Cost of Capital (non-cash) to derive cash only total annual expense)

Fig. A1-1. *Aircraft expenses worksheet.*

AIRCRAFT EXPENSE DATA COLLECTION SHEET

Hangar/Month:		Fuel Cost ($/gal):	
Insurance/yr:		Fuel Cons (gal/hr):	
State Fees/yr:		Oil Cons (qt/hr):	
Annual:		Oil Cost ($/qt):	
Maintenance/yr:		Engine MOH Cost:	
Loan Amount:		Time Rem. To OH:	
Loan/Inv Term-yrs:		Gen Maint Res/hr:	
Loan Interest Rate:		**Aircraft Value:**	
Cost of Capital/yr:		**Com. Rental/hr:**	

Fig. A1-2. *Aircraft expense data collection sheet.*

Piper Archer PA 181   AIRCRAFT PARTNERSHIP FINANCIAL ANALYSIS   ($/Pilot)

NUMBER OF PILOTS	1	2	3	4	5	6

CAPITAL INVESTMENT
LOAN AMOUNT

ANNUAL FIXED EXPENSES
 Tiedown/Hangar
 Insurance
 State Fees
 Annual
 Maintenance
 Loan Payments
 Cost of Capital (non-cash)
 Total Fixed Expenses / yr

HOURLY OPERATING EXPENSES
 Fuel
 Oil
 Engine Reserve
 General Maint Res
 Total Op Exp / hr

TOTAL EXPENSES PER HOUR
 50 Hours flown per year
 100 Hours flown per year
 150 Hours flown per year
 Hourly Commercial Rental

TOTAL ANNUAL EXPENSES
 50 Hours, Own
  Own (cash only)
  Rent
 100 Hours, Own
  Own (cash only)
  Rent
 150 Hours, Own
  Own (cash only)
  Rent

ASSUMPTIONS:

Hangar/Month:		Fuel Cost ($/gal):	
Insurance/yr:		Fuel Cons (gal/hr):	
State Fees/yr:		Oil Cons (qt/hr):	
Annual:		Oil Cost ($/qt):	
Maintenance/yr:		Engine MOH Cost:	
Loan Amount:		Time Rem. To OH:	
Loan/Inv Term-yrs:		Gen Maint Res/hr:	
Loan Interest Rate:		**Aircraft Value:**	
Cost of Capital/yr:		**Com. Rental/hr:**	

Fig. A1-3. *Aircraft and partnership financial analysis spreadsheet layout and sample aircraft financial analysis.*

Piper Archer PA 181   AIRCRAFT PARTNERSHIP FINANCIAL ANALYSIS   ($/Pilot)

NUMBER OF PILOTS	1	2	3	4	5	6
CAPITAL INVESTMENT	50000.00	25000.00	16666.67	12500.00	10000.00	8333.33
LOAN AMOUNT	0.00	0.00	0.00	0.00	0.00	0.00
ANNUAL FIXED EXPENSES						
Tiedown/Hangar	1020.00	510.00	340.00	255.00	204.00	170.00
Insurance	1500.00	750.00	500.00	375.00	300.00	250.00
State Fees	120.00	60.00	40.00	30.00	24.00	20.00
Annual	750.00	375.00	250.00	187.50	150.00	125.00
Maintenance	1000.00	500.00	333.33	250.00	200.00	166.67
Loan Payments	0.00	0.00	0.00	0.00	0.00	0.00
Cost of Capital (non-cash)	3500.00	1750.00	1166.67	875.00	700.00	583.33
Total Fixed Expenses / yr	7890.00	3945.00	2630.00	1972.50	1578.00	1315.00
HOURLY OPERATING EXPENSES						
Fuel	16.00	16.00	16.00	16.00	16.00	16.00
Oil	0.50	0.50	0.50	0.50	0.50	0.50
Engine Reserve	10.00	10.00	10.00	10.00	10.00	10.00
General Maint Res	5.00	5.00	5.00	5.00	5.00	5.00
Total Op Exp / hr	31.50	31.50	31.50	31.50	31.50	31.50
TOTAL EXPENSES PER HOUR						
50 Hours flown per year	189.30	110.40	84.10	70.95	63.06	57.80
100 Hours flown per year	110.40	70.95	57.80	51.23	47.28	44.65
150 Hours flown per year	84.10	57.80	49.03	44.65	42.02	40.27
Hourly Commercial Rental	85.00	85.00	85.00	85.00	85.00	85.00
TOTAL ANNUAL EXPENSES						
50 Hours, Own	9465.00	5520.00	4205.00	3547.50	3153.00	2890.00
Own (cash only)	5965.00	3770.00	3038.33	2672.50	2453.00	2306.67
Rent	4250.00	4250.00	4250.00	4250.00	4250.00	4250.00
100 Hours, Own	11040.00	7095.00	5780.00	5122.50	4728.00	4465.00
Own (cash only)	7540.00	5345.00	4613.33	4247.50	4028.00	3881.67
Rent	8500.00	8500.00	8500.00	8500.00	8500.00	8500.00
150 Hours, Own	12615.00	8670.00	7355.00	6697.50	6303.00	6040.00
Own (cash only)	9115.00	6920.00	6188.33	5822.50	5603.00	5456.67
Rent	12750.00	12750.00	12750.00	12750.00	12750.00	12750.00

ASSUMPTIONS:			
Hangar/Month:	85.00	Fuel Cost ($/gal):	2.00
Insurance/yr:	1500.00	Fuel Cons (gal/hr):	8.00
State Fees/yr:	120.00	Oil Cons (qt/hr):	0.20
Annual:	750.00	Oil Cost ($/qt):	2.50
Maintenance/yr:	1000.00	Engine MOH Cost:	15000.00
Loan Amount:	0.00	Time Rem. To OH:	1500.00
Loan/Inv Term-yrs:	10	Gen Maint Res/hr:	5.00
Loan Interest Rate:	12.00%	**Aircraft Value:**	**50000.00**
Cost of Capital/yr:	7.00%	**Com. Rental/hr:**	**85.00**

Fig. A1-3. *(Continued)*

NUMBER OF PILOTS	1	2
CAPITAL INVESTMENT	=O57-L55	=J6/2
LOAN AMOUNT	=L55	=J7/2

ANNUAL FIXED EXPENSES

Tiedown/Hangar	=L50*12	=J10/2
Insurance	=L51	=J11/2
State Fees	=L52	=J12/2
Annual	=L53	=J13/2
Maintenance	=L54	=J14/2
Loan Payments	=PMT(L57/12,L56*12,-L55)*12	=J15/2
Cost of Capital (non-cash)	=(O57-L55)*L58	=J17/2
Total Fixed Expenses / yr	=SUM(J10:J17)	=SUM(K10:K17)

HOURLY OPERATING EXPENSES

Fuel	=O50*O51	=J22
Oil	=O52*O53	=J23
Engine Reserve	=O54/O55	=J24
General Maint Res	=O56	=J25
Total Op Exp / hr	=SUM(J22:J25)	=SUM(K22:K25)

TOTAL EXPENSES PER HOUR

50 Hours flown per year	=J19/50+J27	=K19/50+K27
100 Hours flown per year	=J19/100+J27	=K19/100+K27
150 Hours flown per year	=J19/150+J27	=K19/150+K27
Hourly Commercial Rental	=O58	=O58

TOTAL ANNUAL EXPENSES

50 Hours,	**Own**	=J30*50	=K30*50
	Own (cash only)	=J37-J17	=K37-K17
	Rent	=J34*50	=K34*50
100 Hours,	**Own**	=J31*100	=K31*100
	Own (cash only)	=J41-J17	=K41-K17
	Rent	=J34*100	=K34*100
150 Hours,	**Own**	=J32*150	=K32*150
	Own (cash only)	=J46-J17	=K46-K17
	Rent	=J34*150	=K34*150

ASSUMPTIONS:	Hangar/Month:
	Insurance/yr:
	State Fees/yr:
	Annual:
	Maintenance/yr:
	Loan Amount:
	Loan/Inv Term-yrs:
	Loan Interest Rate:
	Cost of Capital/yr:

Fig. A1-4. *Aircraft and partnership financial analysis spreadsheet formula template.*

3	4	5	6
=J6/3	=J6/4	=J6/5	=J6/6
=J7/3	=J7/4	=J7/5	=J7/6

=J10/3	=J10/4	=J10/5	=J10/6
=J11/3	=J11/4	=J11/5	=J11/6
=J12/3	=J12/4	=J12/5	=J12/6
=J13/3	=J13/4	=J13/5	=J13/6
=J14/3	=J14/4	=J14/5	=J14/6
=J15/3	=J15/4	=J15/5	=J15/6
=J17/3	=J17/4	=J17/5	=J17/6
=SUM(L10:L17)	=SUM(M10:M17)	=SUM(N10:N17)	=SUM(O10:O17)

=K22	=L22	=M22	=N22
=K23	=L23	=M23	=N23
=K24	=L24	=M24	=N24
=K25	=L25	=M25	=N25
=SUM(L22:L25)	=SUM(M22:M25)	=SUM(N22:N25)	=SUM(O22:O25)

=L19/50+L27	=M19/50+M27	=N19/50+N27	=O19/50+O27
=L19/100+L27	=M19/100+M27	=N19/100+N27	=O19/100+O27
=L19/150+L27	=M19/150+M27	=N19/150+N27	=O19/150+O27
=O58	=O58	=O58	=O58

=L30*50	=M30*50	=N30*50	=O30*50
=L37-L17	=M37-M17	=N37-N17	=O37-O17
=L34*50	=M34*50	=N34*50	=O34*50
=L31*100	=M31*100	=N31*100	=O31*100
=L41-L17	=M41-M17	=N41-N17	=O41-O17
=L34*100	=M34*100	=N34*100	=O34*100
=L32*150	=M32*150	=N32*150	=O32*150
=L46-L17	=M46-M17	=N46-N17	=O46-O17
=L34*150	=M34*150	=N34*150	=O34*150

85	Fuel Cost ($/gal)		2
1500	Fuel Cons (gal/h		8
120	Oil Cons (qt/hr):		=0.2
750	Oil Cost ($/qt):		2.5
1000	Engine MOH Co		15000
0	Time Rem. To O		1500
10	Gen Maint Res/h		5
0.12	**Aircraft Value:**		**50000**
0.07	**Com. Rental/hr:**		85

Fig. A1-4. *(Continued)*

Appendix 2

Sample Agreements
and Administration

Appendix 2.1

SIMPLE CO-OWNERSHIP AGREEMENT, THREE CO-OWNERS

Introduction

The following agreement, entered into by Jane Smith, Joseph Doe, and Donald Roe, hereinafter referred to as the "co-owners," applies to the conduct of their co-ownership in aircraft model Razorback-200, serial number 8, FAA registration number N8BM, hereinafter referred to as the "aircraft."

1. Ownership.

The three co-owners agree that they own three equal, undivided shares of the aircraft.

2. Decisions.

All decisions including decisions to operate away from home base shall be unanimous except as follows:

    2.1. Decisions pertaining to the operation of the aircraft as pilot in command including pre- and postflight activities may be made without the consent of the other co-owners.

    2.2. Scheduling will be on a basis unanimously agreed upon.

3. Legality of operations.

Each co-owner is individually responsible, when pilot in command, to ensure that the operation of the aircraft and related auxiliary activities by the co-owner are within legal and insurance limits.

4. Persons authorized to fly.

No one except the herein mentioned co-owners may be permitted to fly the aircraft, except for test flights by a person who has the appropriate insurance clearance and is unanimously acceptable to all the co-owners.

5. Loan payments.

    5.1. The co-owners agree to share the loan payments under the bank agreement (exhibit A) in three equal parts.

    5.2. The co-owners agree to make all payments by the due date.

    5.3. Each co-owner is individually responsible for charges incurred by his failure to make payments on time.

    5.4. In the event of individual default resulting in foreclosure, the defaulting co-owner agrees to reimburse the other co-owners for all their losses incurred by the foreclosure relative to the original purchase price plus improvements.

6. Expenses.
  6.1. All fixed expenses, as defined in writing by the co-owners from time to time, will be shared equally by the three co-owners.
  6.2. All operating expenses, as defined in writing by the co-owners from time to time, that are incurred by each co-owner, except fuel, will be paid by that co-owner in their entirety.
  6.3. Each co-owner will pay for fuel directly to a fuel supplier of choice and will leave the aircraft fully fueled after each flight.
  6.4. All co-owners will pay their share of expenses to the co-owner designated as Treasurer in a timely fashion. The Treasurer will deposit the payments into the joint co-ownership account and will pay the co-ownership's expenses from that account.
  6.5. In relation to new equipment purchases, said purchases shall not be considered an expense if unanimous agreement to make the purchase is lacking. However, if a co-owner individually decides to purchase a piece of new equipment, he may install it upon unanimous approval, and retains sole rights to it.
  6.6. The co-owners will meet periodically, and in any event at least every six months to review and reconcile accounts.

7. Insurance.
  7.1. Insurance deductibles are to be paid by the pilot in command if the incident/accident is the result of his or her negligence. If individual negligence is indeterminable, the deductible will be shared equally by the co-owners.
  7.2. Insurance receivables in case of incident/accident are to be divided equally among the co-owners or their inheritors, except for 7.1.

8. Rights of related parties.
Parties related to the co-owners possess no rights whatsoever except as follows:
  8.1. In case of a co-owner's death, his inheritors shall be obligated to promptly dispose of his share, with the remaining co-owners having first right of refusal under the terms of the original purchase price plus improvements.

9. Collateralization exclusion.
The co-owners agree not to put up the said aircraft as any form of collateral whatsoever, jointly and severally, except as collateral under the terms of exhibit B.

(Continued)

10. Departure of co-owners from co-ownership.

    10.1. In event of a co-owner's departure from the co-ownership, the remaining co-owner(s) have the first right of refusal for a 45-day period at the original purchase price plus improvements.

    10.2. If a co-owner departs from the co-ownership and the remaining co-owners exercise their option under 10.1, the departed co-owner is to receive a one-third share of any gains on the subsequent sale of the aircraft by the other two co-owners at any time during the first 24 months after the original purchase of the aircraft.

    10.3. If a co-owner departs from the co-ownership and the remaining co-owners do not exercise their option under 10.1, the departing co-owner may sell his share to a third party for any gain, provided that the third party undertakes to unconditionally abide by this agreement.

11. Aircraft value.

Unless the co-owners unanimously agree to an alternate valuation the aircraft value will be the value as defined in the current edition of the *Aircraft Blue Book Price Digest* at the time of valuation.

12. Amendments.

This agreement may be amended at any time upon unanimous agreement of the co-owners. The co-owners are obligated to review the agreement 24 calendar months after its implementation.

Dated this _____ day of _____, 1992.

Witness:_____   _____

                                             Jane Smith

                                             _____

                                             Joseph Doe

                                           _____

                                             Donald Roe

Appendix 2.2

DETAILED CO-OWNERSHIP AGREEMENT, TWO CO-OWNERS

This document memorializes a co-ownership agreement between:

John Smith
of 22 Main Street, Anytown, MA
and
Joe Doe
of 93 Fox Road, Smallville, MA
(referred to hereafter as the "co-owners").

1. The Asset, the co-owners, and ownership.

The co-owners own a 1978 Piper Cherokee Archer aircraft, model PA 28-181, serial number 28-6947032, registered as N4968X (referred to as the "aircraft"). Each co-owner owns an undivided fifty percent (50%) interest in the aircraft.

2. Use of the aircraft.

Each co-owner shall have full use and benefit of the aircraft on a schedule to be mutually agreed upon. No co-owner will use (or permit the use of) the aircraft under conditions which could have the effect of voiding the hull or liability insurance on the aircraft.

3. Legality clause.

All co-owners agree to operate the aircraft in accordance with Federal Aviation Regulations and state and federal laws.

4. Persons authorized to fly.

No person is authorized to fly the co-ownership's aircraft other than the co-owners and appropriately authorized flight instructors providing instruction to co-owners.

5. Decisions.

All decisions except decisions related to the day-to-day operation of the aircraft as pilot in command will be made by consensus.

6. Expenses.

6.1. Fixed expenses.

All fixed expenses arising from the ownership of the aircraft, including but not limited to tie-down or hangar charges, insurance charges, property taxes, and the annual inspection shall be borne equally by the co-owners.

6.2. Operating expenses.

This includes all operating expenses arising from the operation of the aircraft, including but not limited to fuel and other expendables, and maintenance and overhaul reserves. The co-owners agree to

(Continued)

establish mutually acceptable hourly rates for engine overhaul and maintenance reserves at a level that will ensure the accumulation of projected maintenance and engine overhaul costs. The hourly rate for these reserves may be periodically readjusted by consensus as required.

6.3. Payment of expenses except fuel.

The co-owners agree to open a joint co-ownership checking/savings account for making payments related to the aircraft. The co-owners further agree to designate one of their number for a mutually acceptable term as Treasurer, to operate the checking/savings account and maintain a careful record of payment of the aircraft expenses.

Each co-owner agrees to forward payment for his or her share of all fixed expenses and operating expenses, except fuel, in a timely fashion to the Treasurer, who will deposit these payments into the co-ownership account and make payments on behalf of the co-ownership related to the aircraft out of the account.

Except as stated otherwise in this agreement, no expense in excess of $200 may be made on behalf of the co-ownership without prior approval of all co-owners.

The co-owners agree to meet, review, and reconcile all aircraft expenditures at least every three (3) months, when the Treasurer will provide statements for the period covered to all co-owners.

6.4. Assessments.

The co-owners agree to make assessments as necessary to cover unexpected maintenance and engine overhaul expenses if the funds in the account to cover such expenses fall short. Each co-owner will pay a percentage share of the assessment equal to the percentage share of that co-owner's flight time in the total flight time completed in the aircraft by the co-ownership since such maintenance or engine overhaul was last done on the aircraft.

6.5. Payment of fuel expenses.

Each co-owner will pay his or her fuel expenses directly to a fuel supplier of his or her choice. Each co-owner will leave the aircraft refueled to the tabs after every flight.

6.6. Delinquencies.

A co-owner whose payments are 60 days past due loses all flight privileges until the delinquency is cured. A co-owner whose payments are 90 days past due and equal or exceed $2000 may be required at the option of the other co-owners to sell his or her share of the co-ownership to the other co-owners or a third party at a valuation of the annual stated value.

7. Storage.

The aircraft shall be hangared at a airport mutually acceptable to the co-owners.

Away from home the aircraft may be tied down. It is the responsibility of each co-owner using the aircraft to ensure that the aircraft is locked and appropriately secured on the tie-down. When tied down overnight, the aircraft's canopy cover will be put in place.

8. Maintenance and repairs.

The co-owners agree to have the aircraft maintained in accordance with the manufacturer's recommendations and all applicable laws. All maintenance will be appropriately documented. Except for emergency field repairs, all maintenance will be conducted by a maintenance facility chosen by the co-owners' mutual consent.

Major scheduled maintenance and engine overhaul will be scheduled sufficiently in advance to preempt inconveniencing the co-owners.

All maintenance in excess of $500 must be approved by the co-owners' mutual consent.

Co-owners may perform owner maintenance as permitted by FAR 43.3(g), but must have all such work inspected by a mutually acceptable A&P mechanic.

Emergency field repairs away from home may be performed only by an FAA authorized maintenance facility. Such repairs in excess of $500 should first be discussed with the other co-owner, except when a co-owner cannot be reached in spite of all reasonable efforts.

9. Insurance.

The co-owners agree to maintain full hull and liability insurance at a mutually acceptable level to be determined at the beginning of each insurance period and subject to review at any time at the request of a co-owner. The value of the aircraft for insurance purposes will be the value according to the current *Aircraft Blue Book Price Digest.*

10. Loss or damage of the aircraft.

In the event of damage to the aircraft for any reason, the co-owner using the aircraft at the time the damage is incurred will be responsible for payment of the uninsured portion or deductible of the claim. If the aircraft is damaged under circumstances where it was not in use by one of the co-owners, then the uninsured or deductible portion of the claim shall be borne equally.

(Continued)

11. Death or disability of a co-owner.

In the event of the death or disability of a co-owner, the remaining co-owner may purchase the other co-owner's interest for fifty percent (50%) of the most recent agreed-upon fair market value pursuant to paragraph 13 and exhibit A.

12. Departure of a co-owner or termination of the co-ownership.

This co-ownership may be terminated by written notice given to the other co-owner at the address indicated above. Such notice shall also state whether the co-owner electing to terminate desires to buy the aircraft outright or sell his interest to the remaining co-owner. The co-owner receiving such notice shall then inform the other co-owner as to his desire to buy or sell the aircraft. In this situation, the following rules apply:

> 12.1. If both co-owners desire to self the aircraft, it shall be sold to a third party and the proceeds divided equally (following a final accounting between the co-owners).
>
> 12.2. If one co-owner desires to sell and one co-owner desires to buy, the purchase price shall be fifty percent (50%) of the most recent agreed upon fair market value pursuant to paragraph 13 and exhibit A.
>
> 12.3. If one co-owner desires to sell and the other co-owner desires to maintain ownership through a new co-ownership with a third party, the co-owner desiring to sell will give the other co-owner 60 days to find a new co-owner.
>
> 12.4. If both co-owners desire to buy the aircraft, then both co-owners shall submit simultaneously to an independent umpire a sealed bid representing the highest figure that he will pay to purchase the other co-owner's interest in the aircraft. The independent umpire will review both figures, determine which co-owner has offered the higher figure, and the aircraft shall be sold to the co-owner making the higher offer at the amount of that offer.

13. Annual statement of value.

The co-owners agree that as of the date of this agreement, the aircraft's fair market value is $50,000. The co-owners agree that they shall prepare an addendum to this agreement signed by each of them stating the fair market value of the aircraft (to be defined according to a mutually acceptable method) every six months commencing from the date of this agreement in form attached as exhibit A.

14. Collateralization exclusion.

No co-owner may assign, create a security interest in, or pledge his or her interest in the aircraft without the prior consent of the other co-owner.

15. Communications.

Any official request under the terms of this agreement should be made in writing and delivered to the other co-owner by certified mail or private delivery service

16. Arbitration.

In the event that any dispute should arise concerning the co-ownership or the construction of this co-ownership agreement, such dispute shall be submitted to arbitration in accordance with the rules of the American Arbitration Association. The location of such arbitration shall be Anytown, Massachusetts, and the costs of such arbitration proceeding shall be born equally by the parties.

17. Amendments.

This agreement may be amended from time to time by mutual consent of the co-owners.

18. Financing.

The co-ownership has incurred financing for the aircraft as per exhibit B (the loan agreement). The co-owners agree to abide by the terms and conditions of the loan agreement and will each forward their share of the monthly loan payment to the Treasurer by the specified due date. Each co-owner is individually responsible for any charges incurred by his or her failure to make payments on time. In the event of individual default resulting in foreclosure, the defaulting co-owner agrees to reimburse the other co-owner for all losses incurred by the foreclosure.

[alternative language; if co-owner financing has been provided]

The co-owners agree and acknowledge that Mr. Smith's interest in the aircraft has been financed by Mr. Doe pursuant to the terms of a note attached to this agreement as exhibit B. The current principal balance (excluding accrued interest) due on such note is $10,000. Mr. Doe agrees that upon receipt of all future principal payments by Mr. Smith, he will execute an appropriate payments chart appendix modification to this document to reflect the principal payments by Mr. Smith. Mr. Smith acknowledges that his rights in the aircraft and in the aircraft co-ownership are subordinate to the terms of that note and that in the event of loss of the aircraft, death of a co-owner or dissolution of this co-ownership, the note (together with all accrued interest and principal) shall first be paid before he receives any insurance

(Continued)

or sale proceeds. Further, Mr. Smith acknowledges that in the event of any uninsured loss or damage to the aircraft, such uninsured loss or damage shall not serve to reduce or extinguish the note, which will remain as an independent financial obligation.

Dated this _____ day of _____, 1998.

Joe Doe

_____

John Smith

_____

Witness: _____

Appendix 2.3: Sample Accounts

N1786L PARTNERSHIP FINANCIAL STATEMENT 1990

HOURLY RATE

BEGINNING BALANCE

CURRENT BALANCE

INCOME									EXPENSE			

INCOME

Date	Amount	Description	Chk#	Date	Amount	Description	Chk#

EXPENSE

Date	Amount	Payee	Chk#

Fig. A2-3-1. *Simple co-ownership accounts worksheets.*

FINANCIAL SUMMARY 1990

	SZUROVY	SCHUETTE	HOWARD	TOTAL
HOURS				
CASH EXPENSES ($)				
COST OF CAPITAL ($)				
$ PER HOUR				

ENGINE AND AIRFRAME RESERVE

BEGINNING HOURS

ENDING HOURS

BEGINNING E & A RESERVE ($)

ADDITIONAL E& A RESERVE ($)

ASSESSMENTS

ENDING E & A RESERVE ($)

Fig. A2-3-1. (*Continued*).

N1786L PARTNERSHIP FINANCIAL STATEMENT 1990

HOURLY RATE

BEGINNING BALANCE

CURRENT BALANCE

INCOME

Date	Hrs	Op Inc	Oth Inc	Description	Chk#	Date	Hrs	Op Inc	Oth Inc	Description	Chk#	Date	Hrs	Op Inc	Oth Inc	Description	Chk#
		57.60															

Fig. A2-3-2. *Detailed co-ownership accounting worksheet.*

227

N1786L PARTNERSHIP FINANCIAL STATEMENT 1990

Account Types: Hangar, Insurance, Fees, Annual
Maintenance, Op/Res, Other

EXPENSE

Date	Amount	Payee	Chk#	Account

FINANCIAL SUMMARY 1990 ($)

TOTAL

FLIGHT HOURS

HANGAR
INSURANCE
FEES
ANNUAL
MAINTENANCE
OP/RES
OTHER

TOTAL CASH

COST OF CAPITAL

TOTAL EXPENSES

$/HOUR

ENGINE AND AIRFRAME RESERVE

BEGINNING HOURS
ENDING HOURS
BEGINNING E & A RESERVE ($)
ADDITIONAL E & A RESERVE ($)
ASSESSMENTS

ENDING E & A RESERVE ($)

Fig. A2-3-2. (Continued).

Appendix 3

Financing and Insurance

Appendix 3.1: Loan Request Worksheet

	BANK:	BANK:	BANK:
LOAN TERMS Have ready for bankers: Aircraft make, model, year, total time , time since major overhaul, detailed equipment list, damage history, annual date, hours per year to be flown. If specific aircraft is not yet located, define the specs you are looking for.			
Minimum loan amount			
Max % of value financed, new a/c			
Max % of value financed, used a/c			
How value determined?			
Min downpayment requirements			
Max loan maturity, new a/c			
Max loan maturity, used a/c			
Fixed or variable interest rate?			
Annual Percentage Rate (APR)			
Must all partners co-sign for full amount?			
Closing and other fees			
Preapproval available?			
LOAN AMOUNT REQUESTED:			
MATURITY REQUESTED			
MONTHLY PAYMENT			
COLLATERAL REQUIREMENTS			
First lien on aircraft			
Additional collateral requirements			
BORROWER STRENGTH, INFO			
Should each partner be able to carry total loan?			
Application form is sufficient information			
Tax returns also required (how many years?)			
Other evidence of income (W 2, paystubs, etc)			

Fig. A3-1-1. *Aircraft loan request worksheet.*

Appendix 3.2: Aircraft Loan Agreement, Bank Financing

The loan agreement presented here is a typical example of an aircraft bank loan agreement. It is an actual agreement (the names of the borrowers and the bank have been changed) documenting a loan for a Piper Warrior. The loan was fixed rate for 7 years, the maximum maturity this bank was willing to grant for used aircraft. The Warrior's total purchase price was $18,000. Note the terrible "boilerplate" language of the guaranty. It is important to study the loan documents thoroughly, preferably before you show up to sign on the dotted line. Many borrowers make the mistake of not seeing these documents until closing.

FIRST COMMUNITY BANK
CONSUMER AIRCRAFT LOAN
PROMISSORY NOTE, DISCLOSURE STATEMENT
AND SECURITY AGREEMENT

Jane and John Smith

The undersigned, residing at 149 Main Street, Wilcox, Maine; hereinafter called "debtor," hereby grants to First Community Bank, One First Plaza, Anytown, Massachusetts 01390, hereinafter called "secured party," a security interest in the aircraft described below and all accessories and equipment now owned or hereafter acquired by debtor and attached to, located in, or used in connection with such aircraft, including without limitation the engines and avionics equipment described below, all logs and similar books relating to such aircraft, and all proceeds of any of the foregoing, all hereinafter called the "collateral."

Year manufactured:	1976
New or used:	Used
Manufacturer of aircraft:	Piper
Model no.:	Pa28-151
Serial no.:	28-7615026
FAA no.:	N4539X

The avionics equipment includes the following: Dual KX 170B, 2 VOR/LOC, G/S, King Audio Panel, King XPD, King ADF, King DME.

Home airport address: Wilcox Municipal, Wilcox, Maine.

Said security interest is hereby granted as security for the payment of the total of payments hereinafter stated and any substitutions for, or renewals or extension thereof and for the payment and performance of any and all other liabilities and obligations of debtor to secured party under this agreement, all hereinafter called the "obligations."

For value received, debtor agrees and promises to pay to secured party at its principal office in Anytown, Massachusetts, the total of payments ($20,477.52), consisting of the amount financed and the finance charge described below, in consecutive monthly installments in the amounts and at the times described below.

DISCLOSURES OF THE COST OF DEBTOR'S LOAN

Amount financed. The amount of credit provided to debtor or on his behalf:
$12,000.00

Finance charge. The dollar amount the credit will cost debtor:
$8,402.52

Total of payments. The amount debtor will have paid when all scheduled payments have been made:
$20,477.52

Annual percentage rate. The cost of debtor's credit as a yearly rate:
16.50%

Debtor's payment schedule will be:
Number of payments: 84
Amount of each payment: $243.78
When payments are due: monthly, beginning 12/15/98

Late charges: If the amount financed is $6000 or less and debtor's payment is more than 10 days late, debtor will be charged 5% of the overdue installment or $5.00, whichever is less. If the amount financed is over $6000 and debtor's payment is more than 10 days late, debtor will be charged 5% of the overdue installment.

Prepayment: If the obligations are prepaid in full, debtor may be entitled to a refund of part of the finance charge.

Security interest: Debtor is giving a security interest in the aircraft, related equipment, and other assets described above.

Filing fees in connection with the security interest: $52.50

Property insurance may be obtained by debtor through any duly licensed insurance agent of debtor's choice, subject only to the secured party's right to refuse to accept any insurer offered by debtor for reasonable cause.

See the portions of this agreement below for additional information about nonpayment, default, the right to accelerate the maturity of the obligations, and prepayment refunds.

The amount financed is made up of:

Amount paid to debtor directly:	$12,000.00
Amount paid to others on debtor's behalf to public officials:	$22.50
Total:	$12,022.50

ADDITIONAL AGREEMENTS

Debtor further agrees that:

Late payment: If any installment of the total of payments is not paid within 10 days of the due date thereof, debtor agrees to pay secured party a late charge equal to 5% of such overdue installment, which shall not exceed $5.00 if the amount financed hereunder is $6000 or less.

Set-off rights: Any and all deposits or other sums at any time credited by, or due from, secured party to debtor shall at all times constitute additional security for the obligations and may be set off against any obligations upon the occurrence of any event of default, and at any time thereafter.

Prepayment: Prepayment in full of the obligations may be made at any time without penalty. Upon prepayment in full, debtor will receive a rebate of the unearned interest, if any, computed according to the actuarial method, except that no refunds will be made in amounts less than $1 to the extent permitted by law.

Default: If an event of default occurs hereunder, and the secured party declares the balance debtor owes immediately due and payable, debtor will pay interest on that balance (a) for 1 year following the date debtor's balance is declared due and payable at the annual percentage rate set forth above, and from 1 year thereafter at the rate of 6% per annum, until such obligations are fully paid, if the amount financed is $6000 or less or (b) at the annual percentage rate set forth above, until such obligations are fully paid, if the amount financed is over $6000.

Debtor hereby warrants and covenants that, except for the security interest granted hereby, debtor is the owner of the collateral free from any restriction, lien, encumbrance, or right, title, or interest of others, and that debtor will defend the collateral against all claims of all persons at any time claiming any interest therein; that no financing statement covering the collateral or any portion thereof is on file in any public office and that no document covering the collateral and representing a lien, encumbrance, or any right, title, or interest of any party other than debtor or secured party is recorded with the Federal Aviation Administration Aircraft Registry, that debtor will pay all title search, title report, escrow, and filing fees and charges incurred by secured party in connection with the collateral and the perfection of its security interest therein, and that at the request of secured party, debtor will join secured party in executing one or more documents confirming secured party's security interest hereunder, in form satisfactory to secured party, and will pay the cost of recording or filing the same in all public offices whenever filing is deemed by secured party to be necessary or desirable.

DEBTOR UNDERSTANDS AND AGREES THAT THE PROVISIONS APPEARING ON THE REVERSE SIDE HEREOF CONSTITUTE A PART OF THIS AGREEMENT AS FULLY AS IF THEY WERE PRINTED ON THE FACE HEREOF ABOVE DEBTOR'S SIGNATURE. IF THIS AGREEMENT IS SIGNED BY TWO OR MORE DEBTORS, THE TERM "DEBTOR" SHALL REFER TO EACH PERSON SIGNING THIS AGREEMENT, JOINTLY AND SEVERALLY. DEBTOR HEREBY ACKNOWLEDGES RECEIPT OF A COPY OF THIS AGREEMENT.

IN WITNESS WHEREOF, debtor has hereunto set its hand and seal this 10th day of November, 1998:

Witness:

_____   _____
                               Jane Smith (co-owner)

_____   _____
                               John Smith (co-owner)

GUARANTY

For valuable consideration, the receipt and sufficiency of which are hereby acknowledged, the undersigned, jointly and severally if more than one, hereby guarantees to secured party the due payment and fulfillment by debtor of all obligations pursuant to the above promissory note, disclosure statement, and security agreement ("agreement") and agrees that on the default by debtor under the agreement to pay to secured party on demand the full amount unpaid under the agreement. Undersigned consents to any renewal, extension, or postponement of the time of payment of debtor's obligations under the agreement or to any other forbearance or indulgence with respect thereto and consents to any substitution, exchange, modification, or release of any security therefor or the release of any other person primarily or secondarily liable under the agreement whether or not notice thereof shall be given to the undersigned, and the enforcement of secured party's rights hereunder shall not be affected by the neglect or failure of secured party to take any action with respect to any security, right, obligation, endorsement, or guaranty which it may at any time hold. The undersigned waives all requirements of notice (including notice of acceptance of this guaranty), demand, presentment, or protest and any right which the undersigned might otherwise have to require secured party first to proceed against debtor or against any other guarantor or any other person or first to realize on any security held by it before proceeding against the undersigned for the enforcement of this guaranty. An action on this guaranty may be brought in an appropriate court of Massachusetts and service may be had on the undersigned by registered mail.

Date:

Witness:

_____   _____

                               Jane Smith (co-owner)

_____   _____

                               John Smith (co-owner)

[Additional provisions:]

IT IS FURTHER AGREED: Debtor will not sell, transfer, or otherwise dispose of, or offer to sell, transfer, or otherwise dispose of, any or all of the collateral, or any interest therein; that debtor will furnish and keep in force and on deposit with secured party at all times the originals of all risks–ground and flight, liability, and such other insurance policies–as secured party may require, payable to secured party and debtor as their interests may appear, with such supplementary endorsements, coverages, and amounts, and in such form as secured party may require, with insurance companies approved by secured party and noncancelable except on 30 days prior written notice to secured party; that upon failure of debtor to do so, secured party may procure such insurance, and debtor agrees to pay the premiums therefor upon demand, or, if the premiums should have been paid by secured party at its option, to repay the amount thereof to secured party on demand, together with interest thereon at the same rate described herein, but secured party shall be under no duty to procure such insurance or pay such premiums; that insurance proceeds shall be applied toward replacement of the collateral or payment of the obligations at secured party's option, and secured party may act as attorney for debtor in obtaining, adjusting, settling, and canceling such insurance and endorsing any drafts; and that in case of any default hereunder, secured party is hereby authorized to cancel such insurance and debtor hereby assigns to secured party any moneys, not in excess of the unpaid balance of the obligations, payable under such insurance, including return of unearned premiums, and directs any insurance company to make such payments direct to secured party, to be applied to said unpaid balance; that debtor will pay all taxes, assessments, and other charges on or levied against the collateral, will keep the collateral free from any restriction, lien, encumbrance, or right, title, or interest of others and in first-class order and repair and certified for flying at all times, will have the collateral, including all engines and the airframe, thoroughly inspected, overhauled, and repaired as required by the Federal Aviation Administration standards or by the manufacturer and at least once in every period of 12 consecutive months, will report to secured party promptly any damage to the collateral, and will not waste or destroy the collateral or any part thereof or remove any accessories or equipment therefrom; that the collateral will at all times be duly registered with the Federal Aviation Administration and all other federal and state authorities having jurisdiction, that no such registration will at any time expire, or be suspended, revoked, canceled, or terminated, and that the collateral and the use thereof will at all times comply with all laws, rules, regulations, and requirements of the Federal Aviation Administration and all other federal and state authorities having jurisdiction, and the terms and conditions of all said policies of insurance; that secured party may examine and inspect the collateral and all log and similar books and records relating thereto, at any time, wherever located; that the collateral will be used only for the purpose stated above; that the collateral will be kept at all times when not in use for more than 14 days at the home airport address stated herein, and that debtor will not permit removal of any of the collateral from the states to which use of the collateral is restricted by the liability insurance policy required hereunder; that secured party may at its option discharge taxes, assessments, restrictions, liens, encumbrances, rights, title, and interests of others on or in the collateral, and make any reasonable expenditure for maintenance or preservation of the collateral, and debtor will on demand repay the amount thereof to secured party together with interest thereon at the annual percentage rate described above.

The happening of any of the following events, conditions, or occurrences shall constitute an event of default under this agreement: (a) default in the payment or performance of any of the obligations; (b) any warranty, representation, or statement made or furnished to secured party by or on behalf of debtor proves to have been false in any material respect when made or furnished; (c) any loss, theft, damage, destruction, or sale to or of any of the collateral, or the making or suffering of any levy, seizure, or attachment thereof or thereon or the incurring of any restriction, lien, encumbrance, or right, title, or interest of others thereon or therein; or (d) death or insolvency of the debtor, the appointment of a receiver, trustee, or creditor's committee of or for any part of the property of the debtor, the assignment, trust, or mortgage for the benefit of creditors by debtor, or the commencement of any proceeding for a composition of debts or reorganization, or arrangement with creditors, or any proceeding under any federal or state law relating to bankruptcy, insolvency, or the relief or rehabilitation of debtors, by or against debtor.

Upon any such default, which is material under the circumstances and involves either the nonpayment of one or more payments provided for herein or a substantial impairment of the value of the collateral, or at any time thereafter, secured party may, after giving such notice and opportunity to cure as may be required by applicable law, declare the obligations to be immediately due and payable, and exercise, to the extent permitted by applicable law, the rights and remedies of a secured party under the Uniform Commercial Code, including without limitation the sale of the collateral at public or private sale, and any other rights or remedies provided herein. If the secured party sells the collateral at a public or private sale or otherwise disposes of it, and the proceeds of such disposition are insufficient to pay the obligations in full, debtor shall, to the extent permitted by applicable law, remain liable to the secured party for the deficiency with interest thereon at the annual percentage rate described above. If any event of default occurs hereunder, debtor agrees, to the extent permitted by applicable law, to pay secured party's reasonable costs of collection, including court costs and attorney's fees. Any condition or restriction hereinabove imposed with respect to debtor may be waived, modified, or suspended by secured party but only on secured party's prior action in writing and only as so expressed in such writing and not otherwise. Secured party shall not be deemed to have waived any of its other rights hereunder or under any other agreement, instrument, or paper signed by debtor unless such waiver be in writing and signed by secured party. No delay or omission on the part of secured party in exercising any right shall operate as a waiver of such right or any other right. A waiver on any one occasion shall not be construed as a bar to, or waiver of, any right or remedy on any future occasion. All secured party's rights and remedies, whether evidenced hereby or by any other agreement, instrument, or paper, shall be cumulative and may be exercised separately or concurrently. Any demand upon, or notice to, debtor that secured party may elect to give shall be effective when deposited in the mails addressed to debtor at the address shown herein or as modified by any notice given after the date hereof. Demands or notices addressed to any other address at which secured party customarily communicates with debtor shall also be effective. This agreement and all rights and obligations hereunder, including matters of construction, validity, and performance, shall be governed by the laws of Massachusetts. This agreement shall be delivered in Massachusetts by debtor for acceptance by secured party.

If and to the extent that applicable law confers any rights or imposes any duties inconsistent with or in addition to any provisions of this agreement, the affected provisions hereof shall be considered amended to conform thereto, but all other provisions hereof shall remain in full force and effect.

Appendix 3.3: Aircraft Loan Agreement, Partner Financing

The promissory note exhibited here is an example of a loan agreement documenting the financing of one partner by another. To maintain the "arms' length" business-like nature of the transaction, the loan is based entirely on the terms and conditions on which an area bank (the alternative) was providing aircraft financing at the time. The language about how the interest rate is set and varies is similar to what the borrowing partner would have found in a loan agreement required by the bank. The interest rate on this note is variable.

One difference from the bank loan agreement is the percentage point above the "contract interest rate." This percentage is the "spread" the bank charges above a reference rate such as the "contract interest rate" (another common reference rate is the "prime rate"). Had the borrowing partner gone to the bank, he would have had to pay 2.5% over the contract interest rate, resulting in a rate of 11.80%, a full percentage point above the rate of this note. For the lending partner there is also an advantage. The 10.80% the lending partner is making on this transaction is higher than the return on alternative investments, such as certificates of deposit.

PROMISSORY NOTE

$10,000.00   Lincoln, Maine
                                               April 15, 1998

FOR VALUE RECEIVED, the undersigned Peter Piper, promises to pay to Paul Pan the sum of TEN THOUSAND ($10,000.00) DOLLARS with interest from the date of this note on the unpaid principal at the rate of 10.80% per annum, amortized over 120 months, said interest and principal to be paid in consecutive monthly installments of ONE HUNDRED THIRTY SIX AND 75/100 DOLLARS ($136.75), commencing April 15, 1998, until the amount of such consecutive monthly installments is increased or decreased as hereinafter provided. The entire unpaid balance of this note together with accrued interest shall be paid on March 15, 2008.

As of October 15, 1998, the interest rate shall be adjusted to 1.5 percentage points over the "Contract Interest Rate, Purchase of Previously Occupied Homes National Average for All Major Types of Lenders" published by the Federal Home Loan Bank Board or successor agencies, as prevailing 45 days preceding each adjustment date. The interest rate shall thereafter be adjusted in the manner described above every 6 months during the term of this note. In the event of a change in the interest rate, the amount of the monthly principal and interest installment will be changed so that the then unpaid balance of the loan will be completely paid in 120 months from the date of this note.

The entire unpaid balance of this note, together with any interest due thereon, shall become immediately due and payable at the option of the holder hereof upon the happening of any of the following events:

(a) failure of the undersigned to make any payment required to be paid hereunder within thirty (30) days after such payment shall be due; or

(b) in the event that any of the terms, conditions, covenants, or provisions of the mortgage or any other instrument given as collateral security for this note are not fully performed within any applicable grace period; or

(c) upon dissolution or termination of the nominee trust, the trustee of which is the maker of the note herein; or

(d) upon the appointment of a receiver for any part or all of the property of, or an assignment for the benefit of creditors by, or the commencement of any proceedings under any bankruptcy or insolvency laws by or against the maker hereof, other than an involuntary petition which is removed within thirty (30) days.

The maker of this note hereby waives presentment for payment, demand, notice of dishonor, notice of protest, and any other defense, legal or equitable, except payment, which might otherwise be available, and expressly consents to and waives notice of

(a) any extension or postponement of the time for payment or any other indulgence and to the addition or release (whether by operation of law or otherwise) of any other party or person primarily and secondarily liable hereunder; and

(b) any and all impairment, release, substitution, or exchange by the holder hereof of any property securing this obligation.

In the event of any default hereunder, the holder hereof may, at its option, set off against the payment of this note any sums due from the holder to the maker, and may hold, as additional security for the payment of this note, any property, real or personal, of the maker in the possession of the holder.

This note is secured in collateral by a 1978 Piper PA 28-181 Archer aircraft, United States Federal Aviation Administration (hereinafter FAA) registration number N4968X, manufacturers serial number 28-694032. The collateral security agreement shall be a lien placed on the title documents of the aircraft and registered with the FAA, Oklahoma City, Oklahoma.

In the event that the ownership of the aircraft granted as collateral security for this note, or any part thereof, becomes vested in anyone other than the debtor named in said collateral security agreement, the whole sum of principal and interest then remaining unpaid shall become immediately due without notice at the option of the holder hereof.

The maker of this note hereby agrees to pay all costs, charges, and expenses of collection, including reasonable attorney's fees, in the event this note is placed into the hands of any attorney for collection or enforcement hereof.

The rights and obligations hereunder shall be governed by the laws of the State of Massachusetts. In the event that any provision or clause of this note or the collateral security therefor conflicts with applicable law, such conflict shall not affect other provisions of this note or said collateral security which can be given effect without the conflicting provisions, and to this end the provisions of this note and said collateral security are declared severable.

This note has been executed under seal on the day and year first above written

_____   _____
Witness   Peter Piper

Appendix 3.4: Insurance Worksheet

Have for insurers each partner's age, licences and ratings , total hours **logged**, hours in type, hours of relevant experience.

	DESIRED LIMIT	INSURER 1:		INSURER 2:		INSURER 3:	
		COVERAGE	COST	COVERAGE	COST	COVERAGE	COST
HULL							
All risk							
All risk in motion							
All risk not in motion							
Deductibles							
TOTAL HULL COST							
LIABILITY							
Per occurrence							
Per person per occurrence							
Per occupant per occurrence							
Bodily injury sublimit							
Spouse sublimit							
Child (a/c owners') sublimit							
TOTAL LIABILITY COST							
MEDICAL							
Per occurrence							
Per person per occurrence							
TOTAL INSURANCE COST							

EXCLUSIONS (use above insurer numbers for YES/NO)	YES	NO	OTHER TERMS, CONDITIONS AND ISSUES
FAR violations			Flights by non-owners:
Flights on legal airworthiness cert. waivers			Geographic coverage:
Instruction for owners or non-owners			Experience needed to reduce policy cost:
In flight operation by non-owner			Insurance provider rating:
Intentional off airport landings			Other:
Ops under govt. liability waiver			Other:
Other exclusions:			Other:

Note: Only generally undesirable exclusions are listed above. Check policies carefully for other exclusions

Fig. A3-4-1. *Aircraft insurance request worksheet.*

Appendix 4

Aircraft Flight Records

DATE	PILOT	TACH	FLT TIME	ROUTE

Fig. A4-1. *Sample aircraft flight log.*

TACH END:	DATE:		N:		PILOT:		
	ROUTE:						
(-) TACH START:	OAT	ALTITUDE	IAS	MP	RPM	MIXTURE	
= FLIGHT TIME:	OIL PRESS	OIL TEMP	EGT	CHT	GAL/HR	TAS	
REMARKS							

Fig. A4-2. *Aircraft flight and performance log.*

AIRCRAFT SQUAWK REPORT		N:	TACH:
DATE NOTED:	PILOT:	DATE RESOLVED:	BY:
SQUAWK:		RESOLUTION:	
SIGNATURE:		SIGNATURE:	

Fig. A4-3. *Aircraft squawk sheet.*

RECURRENT CHECKS DUE	N:				
ANNUAL INSPECTION:					
100 HOUR INSPECTION:					
TRANSPONDER CHECK:					
STATIC/ALTIMETER CHECK:					
VOR CHECKS:						

Fig. A4-4. *Aircraft schedule of recurrent inspections.*

Appendix 5

Preventive Maintenance

Preventive Maintenance Authorized by FAR 43.3(g)

Part 43 of the Federal Aviation Regulations authorizes the holder of a pilot's license to perform certain preventive maintenance on the aircraft owned or operated by the pilot, provided that the aircraft is not used for part 121, 127, 129, or 135 operations.

It is the sole responsibility of anyone performing any maintenance on any aircraft to be fully informed of and adhere to current maintenance regulations.

Appendix A(c) of FAR part 43 defines preventive maintenance authorized by part 43.3(g) as follows:

- Removal, installation, and repair of landing gear tires.
- Replacing elastic shock absorber cords on landing gear.
- Servicing landing gear shock struts by adding oil, air, or both.
- Servicing landing gear wheel bearings, such as cleaning and greasing.
- Replacing defective safety wiring or cotter keys.
- Lubrication not requiring disassembly other than removal of nonstructural items such as cover plates, cowlings, and fairings.
- Making simple fabric patches not requiring rib stitching or the removal of structural parts or control surfaces. In the case of balloons, the making of small fabric repairs to envelopes (as defined in, and in accordance with, the balloon manufacturer's instructions) not requiring load tape repair or replacement.
- Replenishing hydraulic fluid in the hydraulic reservoir.
- Refinishing decorative coating of fuselage, balloon baskets, wings, tail group surfaces (excluding balanced control surfaces), fairings, cowling, landing gear, cabin, or cockpit interior when removal or disassembly of any primary structure or operating system is not required.
- Applying preservative or protective material to components where no disassembly of any primary structure or operating system is involved and where such coating is not prohibited or is not contrary to good practices.

- Repairing upholstery and decorative furnishings of the cabin, cockpit, or balloon basket interior when the repairing does not require disassembly of any primary structure or operating system or interfere with an operating system or affect the primary structure of the aircraft.

- Making small simple repairs to fairings, nonstructural cover plates, cowlings, and small patches and reinforcements not changing the contour so as to interfere with proper airflow.

- Replacing side windows where that work does not interfere with the structure or any operating system such as controls, electrical equipment, etc.

- Replacing safety belts.

- Replacing seats or seat parts with replacement parts approved for the aircraft not involving disassembly of any primary structure or operating system.

- Troubleshooting and repairing broken circuits in landing light wiring circuits.

- Replacing bulbs, reflectors, and lenses of position and landing lights.

- Replacing wheels and skis where no weight and balance computation is involved.

- Replacing any cowling not requiring removal of the propeller or disconnection of flight controls.

- Replacing or cleaning spark plugs and setting of spark plug gap clearance.

- Replacing any hose connection except hydraulic connections.

- Replacing prefabricated fuel lines.

- Cleaning fuel and oil strainers or filter elements.

- Replacing and servicing batteries.

- Removing and installing glider wings and tail surfaces that are specifically designed for quick removal and installation and when such removal and installation can be accomplished by the pilot.

- Cleaning of balloon burner pilot and main nozzles in accordance with the balloon manufacturer's instructions.

- Replacing or adjustment of nonstructural standard fasteners incidental to operations.

- Removing and installing balloon baskets and burners that are specifically designed for quick removal and installation and when such removal and installation can be accomplished by the pilot, provided that baskets are not interchanged except as provided in the type certificate data sheet for that balloon.

- The installation of antimisfueling devices to reduce the diameter of fuel tank filler openings provided the specific device has been made a part of the aircraft type certificate data by the aircraft manufacturer, the aircraft manufacturer has provided FAA-approved instructions for installation of the specific device, and installation does not involve the disassembly of the existing tank filler opening.

- Removing, checking, and replacing magnetic chip detectors.

Appendix 6

Buying an Aircraft

Appendix 6.1: Aircraft Specification Comparison

	Aircraft (model/yr)	Aircraft (model/yr)	Aircraft (model/yr)	Aircraft (model/yr)
Seats and configuration				
Powerplant make, model				
Horsepower				
Type fuel used (80, 100LL, auto, etc)				
Propeller make and model				
Gross weight				
Empty weight				
Useful load				
Payload, full fuel and oil				
Fuel capacity				
Baggage capacity				
Maximum speed, sea level				
Cruise speed 75% power				
65% power				
55% power				
Range 75% power				
65% power				
55% power				
Rate of climb, sea level				
Service ceiling				
Best angle of climb speed				
Best rate of climb speed				
Stall speed, clean				
Stall speed gear and flaps down				
Approach speed				
Takeoff distance over 50 ft obstacle				
Landing distance over 50 ft obstacle				

Fig. A6-1-1. *Aircraft specifications comparison worksheet.*

Appendix 6.2: Initial Questions to Ask

Aircraft:
Owner:
Telephone:
Address:
Date of conversation:

- How many owners has the airplane had?
- Where has it been based geographically during its life?
- Has it ever been used for training or rental flying?
- Total time airframe?
- Total time engine?
- Total time engine since major overhaul?
- How many times has the engine been overhauled?
- Who performed the overhaul?
- Was the overhaul to factory new tolerances or service limits?
- Were the accessories also overhauled? If not, how many hours are on the accessories (starter, alternator, magnetos, vacuum pump, etc.)?
- Are all ADs complied with and entered into the aircraft logs?
- Has there been a top overhaul since the major overhaul?
- Does the airplane have EGT and CHT gauges?
- Is there any damage history?
- How old is the paint and interior?
- Rate the exterior and interior on a scale of 1 to 10.
- Has the airplane been hangared?
- When was the last annual inspection?
- Who did the last annual inspection?
- Is the oil sent out for analysis at oil change? Are the results available?
- How many hours has the airplane flown since the last annual inspection?
- How many hours has the airplane flown in the last 12 months?
- When was the most recent transponder and static/altimeter check?

- What major maintenance items have been there in the last 12 months?
- What make and model avionics does the airplane have (coms, navs, HSI, ADF, Loran, area nav, GPS, etc.)?
- Are there any maintenance issues with the avionics?
- Does the airplane have an intercom? Does it work?
- Are the airframe and engine logs complete?
- Are pictures available (exterior, interior, and instrument panel)?
- What is the asking price?

Notes:

Appendix 6.3: Prepurchase Inspection Checklist

Aircraft:
Date:
Documentation Inspection

- Airworthiness certificate
- Registration
- Airframe and engine logs. Check for:
  - Proper annual entries
  - Evidence of 100-hour inspections indicating commercial use
  - Compression (most recent and history)
  - Number of hours flown per year
  - Evidence of unscheduled repair work
  - Airworthiness directive compliance
  - Major engine overhaul entry (when and where done, to what tolerances)
  - Record of geographic movements

Mechanical Inspection
Airframe

- Check for wrinkled skin, loose rivets, dings, cracks, and corrosion.
- Check for mismatched paint, which could be a sign of repairs.
- Check all controls for free and correct movement.
- Check all control hinges (ailerons, elevator, rudder, flaps) for looseness, play, and hairline cracks.

- Check vertical and horizontal stabilizer attachment points for looseness, play, and hairline cracks.
- Check wing attachment points for hairline cracks and corrosion (there better not be any looseness).
- Check control cables for looseness and chafing. Look inside fuselage and wings through inspection panels with a flashlight.
- Check fuel caps, quick drains, and fuel tank areas for signs of fuel leaks (brownish stains).
- Check fuselage underside for cleanliness and signs of leaks from engine area.
- Check wing struts for any signs of damage, corrosion, or hairline cracks.
- On fabric covered aircraft do fabric test and check for loose or peeling fabric.
- Check engine cowling for looseness, play, and cracks, especially at attachment points.

Landing Gear

- Check landing gear struts for leaks.
- Check tires for wear and bald spots.
- Check brake pads for wear, brake disks for corrosion, pitting, and warping, and brake hydraulic lines for signs of seeping or leaking fluid.

Cockpit

- Check cabin doors and windows for signs of water leaks.
- Check windows for crazing.
- Check seat belts for wear and tear. They can get caught and damaged in the seat rails.
- Move all controls and trim to verify full control movement and check for binding.
- Move all other knobs and switches to check for proper operation.
- Check ELT for proper operation.
- Check entire aircraft for proper display of placards and limitations.

Engine and Propeller

- Check compression.
- Check baffles for damage or deformation. Baffle irregularities can cause cooling problems.

- Check for any sign of leaks, especially around the various gaskets. Look for oil and fuel stains.
- Check lower spark plugs for proper condition (take them out and examine them).
- Check wiring harness for signs of brittleness and fraying.
- Check the induction/exhaust system for leaks, cracks, corrosion, and looseness.
- Check engine controls running to the cockpit for free and easy movement.
- Check the battery fluid level and for signs of overheating.
- Check accessory attachments (alternator, magnetos, vacuum pump, starter, electric fuel pump, etc.). Check alternator belt for fraying and tightness.
- Check propeller for nicks and spinner for cracks.

Appendix 6.4: Flight Test Checklist

Aircraft:
Date:

- Preflight. Do a detailed preflight per the operating manual. Follow the detailed manual checklist to be sure of covering everything.
- Engine start. Note how easily the engine cranks and turns over.
- Taxi. Perform radio checks on both coms. Test brakes. Notice engine response, steering, and suspension.
- Pretakeoff checks. Perform careful pretakeoff checks and run-up, verifying aircraft behavior to operating manual standards.
- Takeoff. Note if power settings reach required levels. Climb to a convenient maneuvering altitude.
- Basic VFR maneuvers, handling checks. Perform steep turns, slow flight, and stalls. Cycle flaps and gear. Note if aircraft is in trim.
- Cruise checks. Set up various cruise configurations. Note performance compared to operating manual standards. Check engine gauges.

- Avionics checks. Check all avionics in flight. Do in-flight radio checks. Track VORs, localizer, and glide slope on all appropriate navigation instruments. Compare needle indication of each VOR set to the same radial. Check ADF and marker beacons as appropriate. Establish radar contact with ground control. Compare transponder altitude readout reported by ground control to altimeter indication.

- Miscellaneous checks. Check cabin lights, vents, and heating system.

- Touch and goes. Perform two or three touch and goes to feel fully comfortable with the handling of the aircraft.

- Postflight check. Uncowl engine, check for evidence of any leaks. Turn on all navigation lights, landing lights, strobes, and flashing beacon, and check for operation.

Appendix 6.5: Closing Checklist

Aircraft:
Date:
You get from the seller

- signed aircraft bill of sale
- airworthiness certificate
- airframe and engine logs
- aircraft operating manual
- weight and balance certificates

You give the seller

- funds in the amount of the sales price (minus the deposit already given).
- copy of the bill of sale for the seller's records. This is optional but most sellers will want one (you will have to make a copy because the FAA form is only in duplicate, the original to be sent in, the copy to be kept by you).

Place in the aircraft to meet FAA regulations

- airworthiness certificate
- new temporary registration
- operating manual
- weight and balance certificates

Appendix 6.6: Sample Documents

FORM APPROVED
OMB NO. 2120-0042

UNITED STATES OF AMERICA
DEPARTMENT OF TRANSPORTATION FEDERAL AVIATION ADMINISTRATION

AIRCRAFT BILL OF SALE

FOR AND IN CONSIDERATION OF $ THE UNDERSIGNED OWNER(S) OF THE FULL LEGAL AND BENEFICIAL TITLE OF THE AIRCRAFT DES-CRIBED AS FOLLOWS:

UNITED STATES REGISTRATION NUMBER N

AIRCRAFT MANUFACTURER & MODEL

AIRCRAFT SERIAL No.

DOES THIS   DAY OF   19

HEREBY SELL, GRANT, TRANSFER AND DELIVER ALL RIGHTS, TITLE, AND INTERESTS IN AND TO SUCH AIRCRAFT UNTO:

Do Not Write In This Block
FOR FAA USE ONLY

PURCHASER

NAME AND ADDRESS
(IF INDIVIDUAL(S), GIVE LAST NAME, FIRST NAME, AND MIDDLE INITIAL.)

DEALER CERTIFICATE NUMBER

AND TO   EXECUTORS, ADMINISTRATORS, AND ASSIGNS TO HAVE AND TO HOLD SINGULARLY THE SAID AIRCRAFT FOREVER, AND WARRANTS THE TITLE THEREOF.

IN TESTIMONY WHEREOF   HAVE SET   HAND AND SEAL THIS   DAY OF   19

SELLER

NAME (S) OF SELLER (TYPED OR PRINTED)	SIGNATURE (S) (IN INK) (IF EXECUTED FOR CO-OWNERSHIP, ALL MUST SIGN.)	TITLE (TYPED OR PRINTED)

ACKNOWLEDGMENT (NOT REQUIRED FOR PURPOSES OF FAA RECORDING: HOWEVER, MAY BE REQUIRED BY LOCAL LAW FOR VALIDITY OF THE INSTRUMENT.)

ORIGINAL: TO FAA

AC FORM 8050-2 (8-85) (0052-00-629-0002)

Fig. A6-6-1. *FAA bill of sale.*

FORM APPROVED
OMB NO. 2120-0029
EXP. DATE 10/31/84

UNITED STATES OF AMERICA DEPARTMENT OF TRANSPORTATION
FEDERAL AVIATION ADMINISTRATION-MIKE MONRONEY AERONAUTICAL CENTER
AIRCRAFT REGISTRATION APPLICATION

CERT. ISSUE DATE

UNITED STATES
REGISTRATION NUMBER **N**

AIRCRAFT MANUFACTURER & MODEL

AIRCRAFT SERIAL No.

FOR FAA USE ONLY

TYPE OF REGISTRATION (Check one box)

☐ 1. Individual ☐ 2. Partnership ☐ 3. Corporation ☐ 4. Co-owner ☐ 5. Gov't. ☐ 8. Foreign-owned Corporation

NAME OF APPLICANT (Person(s) shown on evidence of ownership. If individual, give last name, first name, and middle initial.)

TELEPHONE NUMBER: (   )   –

ADDRESS (Permanent mailing address for first applicant listed.)

Number and street: _____

Rural Route: _____   P.O. Box: _____

CITY	STATE	ZIP CODE

☐ **CHECK HERE IF YOU ARE ONLY REPORTING A CHANGE OF ADDRESS**

ATTENTION! Read the following statement before signing this application.

A false or dishonest answer to any question in this application may be grounds for punishment by fine and / or imprisonment (U.S. Code, Title 18, Sec. 1001).

CERTIFICATION

I/WE CERTIFY:

(1) That the above aircraft is owned by the undersigned applicant, who is a citizen (including corporations) of the United States.

(For voting trust, give name of trustee: _____), or:

CHECK ONE AS APPROPRIATE:

a. ☐ A resident alien, with alien registration (Form 1-151 or Form 1-551) No. _____

b. ☐ A foreign-owned corporation organized and doing business under the laws of (state or possession) _____, and said aircraft is based and primarily used in the United States. Records of flight hours are available for inspection at _____

(2) That the aircraft is not registered under the laws of any foreign country; and
(3) That legal evidence of ownership is attached or has been filed with the Federal Aviation Administration.

NOTE: If executed for co-ownership all applicants must sign. Use reverse side if necessary.

TYPE OR PRINT NAME BELOW SIGNATURE

EACH PART OF THIS APPLICATION MUST BE SIGNED IN INK.	SIGNATURE	TITLE	DATE
	SIGNATURE	TITLE	DATE
	SIGNATURE	TITLE	DATE

NOTE: Pending receipt of the Certificate of Aircraft Registration, the aircraft may be operated for a period not in excess of 90 days, during which time the PINK copy of this application must be carried in the aircraft.

AC FORM 8050-1 (1-83) (0052-00-628-9005)

Fig. A6-6-2. *FAA certificate of registration.*

Index

About the Author

Geza Szurovy (Boston, MA) is an experienced aviation writer who has co-owned several personal airplanes in aircraft partnerships. He is the author of seven McGraw-Hill books, two of which won Aviation/Space Writers' Association Awards of Excellence. Szurovy is a commercial pilot with instrument, multiengine, DC-3, and glider ratings. He has flown and written about a wide variety of aircraft, from Piper Cubs to Learjets.